WITHDRAWN

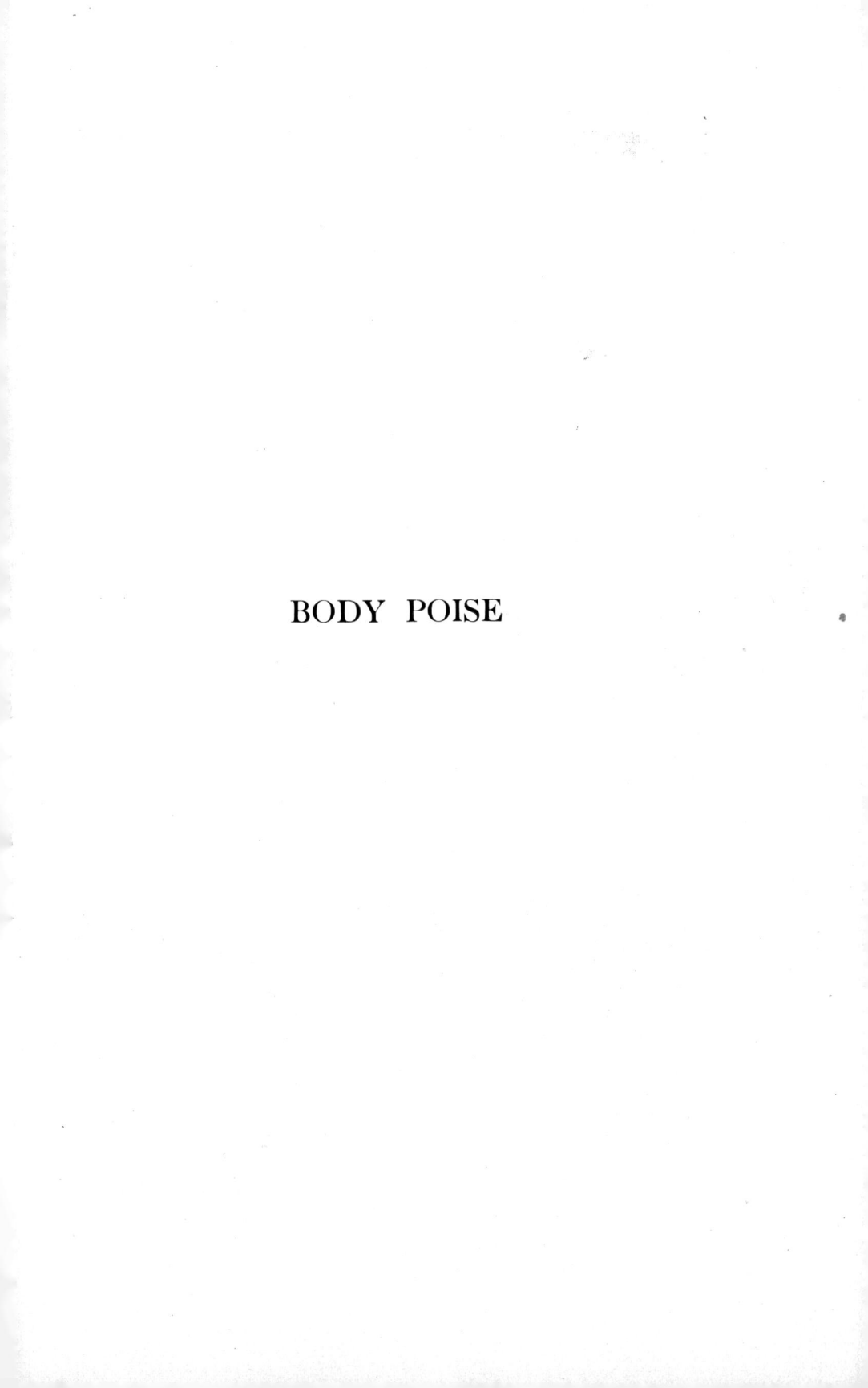

BODY POISE

DAVID BY MICHAEL ANGELO
(Photograph from original in the Academy of Fine Arts, Florence, Italy—1475-1564)

BODY POISE

By

WALTER TRUSLOW, M.D., F.A.C.S.

Consulting Orthopedic Surgeon to the following hospitals: Brooklyn, Long Island College, St. John's, Norwegian, Victory Memorial, Evangelical Deaconess, Kingston Avenue, St. Giles, Brooklyn, N. Y. and Pilgrim State, Brentwood, N. Y.; former Lecturer on Orthopedic Surgery, Long Island Medical College and on Anatomy and Kinesiology, New York University, New York School of Physical Education and Y.M.C.A. Summer School, Silver Bay, N. Y.

THE WILLIAMS & WILKINS COMPANY

BALTIMORE

1943

Copyright, 1943
The Williams & Wilkins Company

Made in the United States of America

Published 1943

Composed and Printed at the
WAVERLY PRESS, INC.
FOR
The Williams & Wilkins Company
Baltimore, Md., U. S. A.

*"There is no theam more plentifull to scan
Than is the glorious goodly frame of man"*

(DU BARTAS 1544–1590)

Grateful acknowledgement is made to Dr. Robert L. Dickinson for encouragement in presenting this book and for aid in editing the illustrations; to Dr. John B. Truslow, for suggestions in the text; and to the Orthopedic and Photographic Departments of Brooklyn Hospital and Long Island College Hospital in supplying some of the models and the reproductions of many of the photographs.

FOREWORD

Body poise has always held a leading place in physical fitness, and should continue to loom large in our national program. Nice balance from head to feet not only lessens strain of the structures involved but, by maintaining vital organs in their normal relationships, each to each, enhances efficient functioning of these organs. Thus, well balanced feet minimize foot and ankle strains and affect all segments and joints above the feet; good balance at the knees, the hip and the low spine prevent many of the strains and aches of these regions; the maintenance of the normal physiological curves of the spine and of lateral symmetry of the trunk tends to avoid the stresses and local pains of backache, and the well poised head and neck prevent many of the high spine discomforts and even headaches. Of equal importance is the effect of posture on the vital organs of the chest, the abdominal and pelvic cavities. So, lacking infectious and traumatic causes, good posture enhances normal functioning of the lungs and of the heart, of the stomach, the intestines and the auxiliary organs of digestion, of the kidneys, the bladder, and of the entire genito-urinary system. In this way the long train of deleterious effects of organic displacements and faulty pressures, which are in large measure directly traceable to postural mal-positions, are prevented. It must therefore be evident that good body poise is an essential to physical fitness.

As a nation we have lagged behind the totalitarians in body training of our youth. In the latter part of the last century, more or less adequate physical training had an important place in the programs of our primary and secondary schools. Well trained instructors often occupied honored places on school and college faculties. At that time, much attention was given to good posture and to rational training for physical efficiency, leaving it for the few to attain specialized training for athletics and for the performance of gymnastic "stunts." Since then school-yard games, rhythmic dancing and more general participation in athletics, valuable as they are, have gradually superseded the scientific posture training and body-building programs of the Swedish "day's order" or of modifications of this and of "setting-up" drills. This has resulted in a real let-down in basic preparation for life. But the War, and especially our participation, is already bringing about a change in our attitude towards physical training. This change is beginning in our elementary schools, where more and more time in the day's program of studies is allotted to setting-up drills and scientific body building work. Our daily press tells of the replanning of the schedules, of increasing numbers of city and town Boards of Education, to this end. The thousands of new entrants into the Army, the Navy, the Marine and the Air Service will, of course, get much ground-work physical training. If this awakening attitude should spread to industry—both for men and for women, for defense and for non-defence—we may well look toward a people as truly physically fit as we like to believe ourselves morally and industrially fit.

To set forth the mechanism of good posture is the thesis of the first part of this book. The second part deals with the restoration of the lapses from good posture. Part III analyzes body poise in relation to sports and games.

CONTENTS

Introduction and General Principles.................................... 1

Part I
THE HUMAN FRAME—ANATOMY AND KINESIOLOGY

Introduction.. 8
Chapter 1. Fundamental Standing Position.......................... 13
Chapter 2. Other Fundamental Gymnastic Positions................ 125

Part II
SOME PATHOLOGICAL DEVIATIONS FROM THE NORMAL INVOLVING POSTURE MECHANISM AND TREATMENT

Introduction.. 136
Chapter 3. Weak Feet, Flat Feet, Metatarsalgia..................... 138
Chapter 4. Faulty Posture—Functional Antero-Posterior and Functional Lateral Deviations of the Spine................... 167
Chapter 5. Rotary Lateral Curvature of the Spine—Scoliosis........ 199

Part III
BODY POISE IN GAMES, SPORTS AND ATHLETICS

Introduction.. 239
Chapter 6. Sports and Contests Presenting Little or No Tendencies to Poor Posture... 242
Chapter 7. Sports and Games Having Mild Tendency to Faulty Body Poise.. 255
Chapter 8. Sports and Games with Greater Tendency to Horizontal Disproportion of Muscular and Structural Development and Lateral Asymmetry............................. 273

Glossary of Anatomical Terms.. 286

Index.. 299

INTRODUCTION AND GENERAL PRINCIPLES

In the history of medicine, the consideration of body mechanics forms one of the corner stones of our knowledge of normal and of abnormal functions of the human system. A nice balance of many superimposed irregularly shaped segments of bone, held together by strong ligaments and muscles and motivated by muscles, which are controlled by the nervous system, is the picture of good body poise. The more or less habitual and, in some regions, occasionally fixed malalignment of these superimposed segments, with disturbance of the anatomy and physiology of the controlling structures, usually leads to poor body poise. The regaining of well-balanced alignment or at least compensating relationships of the parts, with return of healthful function of the controlling members, is the aim of treatment looking toward the restoration of good body poise. Systematic gymnastics plays a leading role in the maintenance of good posture and corrective gymnastics in the restoration of good body poise. This is the theme of this book.

So we study the skeletal bones, the ligaments which hold them together and the muscles which activate them. Muscles require normal activity for the preservation of their anatomical and functional integrity. Many of the commonest physical affections of man are largely referable to simple muscle imbalance or insufficiency. Swedish physicians of the early part of the last century pioneered in the search and research for correlation between the errors in bodily mechanics, which are so often associated with pain and disability, and the errors in man's habits which predispose to these muscular insufficiencies in every day life. It has long been known that accident and disease also tended toward muscle imbalance. There has gradually been developed a sensible appreciation of the *prophylactic* effectivenes of the fundamentals of good posture and of the *therapeutic* value of corrective gymnastics for conditions of widely diverse complaint.

Corrective exercises have long been used in the care of human ailments. It is therefore fitting to gather into one manual certain physical conditions in which *muscular insufficiency plays a leading part*. This is strikingly true of weak feet, of faulty posture and of lateral curvature of the spine. These three forms of body derangement are also so interrelated that it is desirable to discuss them together.

Part I, of this book, treats of the normal anatomy and the kinesiology—the bone, joint and muscle relationships—in good posture. The anatomy is presented as an "up building", from soles of the feet to the crown of the head, of the structures necessary for maintaining and using the human frame, in normal posture. This differs from the regional plan of the classical text books on anatomy. Under kinesiology, this text will explore occupational and gymnastic variations from the normal posture and the changes in the mechanism which produce these variations. Herein are studied the anatomy, kinesiology and leverage principles of fundamental gymnastic positions.

Were we to examine the human skeleton as a piece of mechanical engineering—as a contractor eyes a blueprint—we might be inclined to reject it. Simply impracticable, could be the verdict. Like the Palace of the Doges in Venice, it would be found beautiful but defiant of sound building principles. That great edifice is now webbed with steel bracing in order to maintain its integrity, even at rest. The human frame walks and runs—and stops walking and running! And the human voice cries out when the walls buckle a bit, the foundations sink a little, the human columns sag.

Mankind cannot condemn as unsafe or unfit this temple of his noble mind. But mankind can use his mind far more effectively than he does, to examine the opportunities for more poised maintenance and practical repair of the habitus which is his by immutable legacy.

Delicate foundation stones are the *feet*. Well poised and controlled they must be to distribute, to the best advantage, over the os calcis, the tarsus and the heads of the metatarsal bones, the onus of a hundred-odd pounds. Forwardly directed toes and adequately supported medial arching of the graceful plantar architecture are essential to attainment of the greatest possible mechanical efficiency.

Upon these foundations, the central column of the *spine* must graduate its weight equally; and for this, the Gothic arch of legs and pelvis is admirably adaptable. But the column atop the arch is straight only from the antero-posterior viewpoint; while from a lateral aspect, the curves involve not only the spine, but the pelvic arch itself. (see Fig. 2, p. 14) Even if all of the bony ends could be squared horizontally, it is clear that there is no conceivable position in which the human skeleton could be maintained in a perpendicular balance, without extensive bracing. In life, this is provided by ligaments and muscles; and it is important to note that, except in lying positions, there is no possible approach to a symmetrical relaxation of these sinuous stays. In other words, every position assumed by the human body—except possibly flat in dorsal or ventral recumbency—is accompanied by asymmetrical stress upon its supporting elastic elements.

Fortunately for us, the human mechanism is endowed, in health, with innate power of recovery from these faulty tendencies, and automatically rights itself with cessation of the harmful factors.

Chapter I fully describes these body relationships in fundamental standing position. Chapter 2 portrays seven other fundamental gymnastic positions, which are useful in corrective gymnastics. These other gymnastic positions are studied in terms of variation from Fundamental Standing Position.

Although inadequacy of muscle development and of muscle training constitute leading factors in each of the derangements discussed in Part II and although muscle re-education is a potent means in restoring these parts to normal, there are *other factors* and *other methods* of *treatment* which should not be neglected. Rightly to understand the value of exercise treatment, the operator, whether orthopedic surgeon, general practitioner of medicine, physiotherapist, gymnasium director, masseur or nurse, should be well informed concerning these other means used in the correction of the disabilities discussed. He should have a clear picture of the entire care of the patient. So this manual outlines the entire treatment for the conditions discussed, and states the place of corrective gymnastics in the total.

Statistically, the largest group of patients to be adequately benefited by muscle re-education are the sufferers from some form of faulty body posture, both in action and at rest.

* * * * *

For the sake of clearness, weak feet, faulty posture and lateral curvature of the spine will be considered in separate chapters in Part II. In the discussion of each, particular emphasis will be placed upon:

Definition with special reference to physical deviations;

Accuracy of *Measurement* and *Recording* of the deformity—with presentation of the author's simple and generally applicable methods;

Outline of the *Entire Treatment*—with reference to all of the usual measures which are useful;

Tabulation, Description and Illustration of certain effective *Exercises;* and

Evaluation of the relative value of corrective exercises to the entire treatment.

This manual will not specifically tabulate symptoms. They are efficiently presented by many writers.

Estimate of the patient's present condition and of progress during treatment is largely dependent upon *measurement* and *record making*. The simpler the methods employed, the more useful. Those presented here meet this need. X-ray examination is of importance, especially in scoliosis, but need not be made as frequently as the simpler follow-through measurements here presented.

We all know the inadequacy of having at our disposal but one method of treatment. We appreciate the menace of those irregular practitioners who have but "one string to their bow". The gymnastic enthusiast may lay stress upon exercises to the exclusion of other necessary treatment, especially to that included under the word Protection. On the other hand, the medical practitioner may give too little heed to muscle development work, which alone can fully restore the part involved in the disability. So, in the discussion of these three bodily derangements, the *entire treatment* will be given.

Actually, the *raison d'etre* of Part II, of this manual, is to emphasize and to gather together, in an orderly form, certain exercises which are useful for each of the conditions discussed.

Of many postures and movements available, emphasis is herein laid upon those which have been found most effective in the writer's experience of many years. The original exercises for rotary lateral curvature of the spine, although recording a radical departure from orthodox use, have been found so uniformly effective that they are herein presented almost to the exclusion of others. Also some of the movements used for weak feet, flat feet and metatarsalgia, depart from the standing position of the text books, although many of the exercises are similar to those often used in standing position. Most of the corrective work for feet will be given as in the sitting position, because only thereby can be attained greater accuracy of movement and less strain upon weakened parts.

All of the exercises appearing in this manual require but little, if any, gymnastic apparatus. They are therefore suitable for use in a physician's treatment room as well as in the equipped gymnasium; and they may be practiced, after initial training under supervision, by the patient in his home. Although individual muscle training usually brings best results, where adequate floor space is available, these exercises may be given in classes not to exceed eight or ten in number of persons. For some ages of patients, the group method of instruction has a real advantage by furthering mutual stimulation to effort; but the leader of such a group, whether physician or one non-medically trained, must be vigilant to insure good "form". *Accuracy of performance of these exercises and doing them to the full extent of the patient's present ability constitute a large part of their effectiveness.*

As a practical procedure, the writer has found that during a course of treatment of these patients, it is of the utmost importance to evaluate well the status of the musculature involved, at frequent intervals. For on this estimation depends the determination of the amount of Protection indicated—protection from muscle and ligament strain and from structural deformity. And also upon this estimation depends the extent of active Muscle Re-education both safe and advantageous. The following diagram well illustrates the relative status of protection and of muscle re-education in these disabilities (as indeed it does in many other orthopedic conditions):

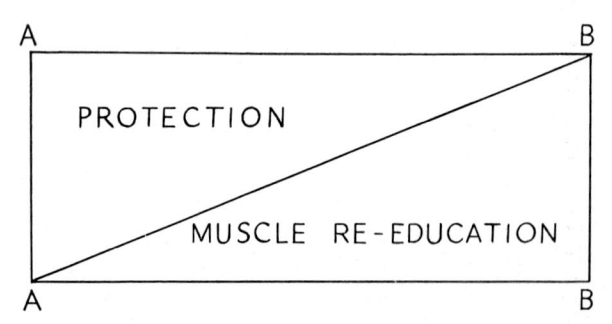

```
A                                    B
┌─────────────────────────────────────┐
│   PROTECTION          ╱             │
│              ╱                      │
│       ╱         MUSCLE  RE-EDUCATION│
└─────────────────────────────────────┘
A                                    B
```

The line A–A, expresses the beginning of treatment. Protection is all important; muscle re-education begins gradually. The line B–B is that status in which muscle self-sufficiency has been developed enough to warrant the discarding of protection. Between these two there is a gradual increase in muscle training and in the attainment of muscle self-sufficiency and a gradual decrease in the necessity for the use of protection.

This inter-relationship is dwelt upon because neglect of one or over-eagerness in using the other means of management, has time and again unnecessarily prolonged the course of therapy, turned normally expectant progress into retreat and even actively predisposed to indefensible complications of over-emphasized treatment.

Some leading sports and games, in their relationship to body poise, are presented in Part III. Six such are defined, with short historical sketch, with full kinesiological analysis, with notation as to their relationships to body poise and with suggestions concerning methods of counteracting faulty tendencies. Twenty-one other sports are included, with brief comment, but without analysis. All sports and games, named in Part III, are classified as to their relationships to body poise—that is, as having little or no faulty tendency, as involving mild tendency and finally those showing more risk to symmetric body development and requiring counteracting work. The neutralizing procedures are presented, in some cases, as symmetrical gymnasium work, sometimes as other sports and, in a few events, as corrective exercises found in Part II.

It is recognized that the aim of some athletic coaches may be toward the training and perhaps overtraining of specialized muscles and muscle co-ordinations to the possible detriment of the best upbuilding of the bodies of the participants. But there are not many such coaches. On the other hand, many coaches and certainly the majority of physical trainers today recognize that *basic, all-around body training* is the *best foundation* for *excellence* in *athletic results*, and that the *specialized training* of *individuals* or of the members of a *team* should be toward *co-ordination* of *body efforts* to attain the ends sought. The work of the physical director should precede that of and should follow-through with that of the coach.

This treatise will not specifically detail basic physical training for specialized athletic events. It is well covered by others. But it is believed that there is real need for more study of the physical technique (kinesiology) of sports, with authoritative emphasis on those that are beneficial and caution as to the use of those which tend toward uneven and asymmetric physical development—and with suggestion as to counteracting bad trends. That is the purpose of this section.

In this book, illustrations have been liberally used. For the reader, not fully trained in his anatomy, a glossary of anatomical terms is appended. This is followed by a general index.

Part I

THE HUMAN FRAME—ANATOMY AND KINESIOLOGY

INTRODUCTION

The human body may be considered, in the standing position, as an architectural structure. It may be described as a tower, with a base, a body and a super structure. The base, which is bifundamental, is comprised of the feet, the legs and the thighs; the body is the trunk, including the pelvis, the lumbar and thoracic spine and the chest; the super structure is the neck, the head and the arms. The principal reliance for the maintenance of this architectural structure is found in the many bones of the human skeleton; the auxiliary factors, in this human tower are the ligaments, which hold the bones together, and the muscles in tonic contraction. This is the Anatomy.

But the human body is much more than a tower. It is an animated tower—a movable machine. All of the bones (except those of the skull and the pelvis) may be moved each upon or in relation to its neighbor, at the body joints; ligaments prevent contiguous bones from actual separation but allow much motion between them, and muscles, controlled by the motor side of the central nervous system, initiate and regulate the movements of the many parts of the framework. In health, bodily movements are purposeful and co-ordinate. We possess a wonderful piece of architecture; we are endowed with a very useful machine. The study of this frame work in motion is called Human Kinesiology.

The discussion of this Anatomy and this Kinesiology is the purpose of Chapter 1 of this book. The anatomical approach differs from that of usual text books on the subject. As Corrective Gymnastics is a leading theme of this book, certain Fundamental Gymnastic Positions will be defined and portrayed. Fundamental Standing Position, embodied in Chapter 1, is the basis of the study of the bones, the joints and the muscle controls. The Position will be "built up" from the ground base to the head and the arms, segment by segment, each being completed before the next one is discussed. The study of the Kinesiology of the parts will be co-ordinated with the Anatomy, segment by segment. Chapter 2 defines seven other gymnastic "starting positions".

Kinesiology is the science of motion. It includes all kinds of motion. *Animal Kinesiology* is the science of motion in the animal kingdom. *Human Kinesiology* is the science of motion in the human being. *Gymnastic Kinesiology* is the science of motion as applied to gymnastic positions, postures and exercises. *Corrective Gymnastic Kinesiology* is the science of motion as applied to remedial positions, postures and exercises.

Gymnastic Kinesiology and Corrective Gymnastic Kinesiology will be discussed in this book.

The *Fundamental Gymnastic Positions* used in this manual are:
1. Standing;
2. Kneeling;
3. On Hands and Knees;
4. Lying;
5. Prone Lying;
6. Half-Prone Lying at the End of Table;
7. Sitting;
8. Spring-Sitting.

All of these have the *center of gravity* above the base of support and the *line of gravity* passing vertically from the center of gravity to some point within the base of support (see Fig. 2, p. 14). In all of these positions, there is stable equilibrium. In the exercises derived from the fundamental gymnastic positions, the center of gravity may change but the line of gravity passes downward to a point, usually still within the base of support, but other than that of the starting position.

The fundamental gymnastic positions are defined by body contacts with fixed supports. The *base of support* is that area on the floor, or other supporting surface, which is outlined by the extreme points of contact, thus:
1. The Standing Position is defined by foot contacts (see Fig. 2, p. 14),

2. The Kneeling Position is defined by knee and toe contacts (see Fig. 26, p. 127),
3. The Position On Hands and Knees is defined by contacts of palms of hands and of knees and of backs of the feet (see Fig. 28, p. 129),
4. The Lying Position is defined by contacts of the rear of the head, the shoulders, the trunk, the buttocks, the thighs, the legs and the heels (see Fig. 29, p. 131),
5. The Prone-Lying Position is defined by entire front body contacts (see Fig. 30, p. 131),
6. The Half-Prone Lying Position is defined by contacts of the front of the head, the trunk (on the table) and the fore parts of the feet (on the floor) (see Fig. 31, p. 132),
7. The Sitting Position is defined by buttock and posterior thigh contacts (on the bench) and by contacts of the soles of the feet (on the floor) (see Fig. 32, p. 133),
8. The Spring-Sitting Position is defined by posterior thigh and buttock, of one side (on a bench), by the sole of the foot, of the same side (on the floor), by the tips of the extended toes of the opposite side (on the floor) (see Fig. 33, p. 134).

The principles of *Leverage* are ever recurring factors in the study of Body Mechanics. Physicists have named three elements of the lever and have divided levers into three orders. The elements are:

a. Power (P);
b. Fulcrum (F);
c. Weight or Resistance (W).

The orders of leverage are: (see Fig. 1)

a. Lever of the first order, PFW, in which the fulcrum is found between the power and the weight;
b. Lever of the second order, FWP, in which the weight is found between the fulcrum and the power; and
c. Lever of the third order, FPW, in which the power is found between the fulcrum and the weight.

All three orders are found in the human mechanism, but the Lever of the third order, FPW, is the most common.

To measure the effectiveness of leverage activity, in all animal mechanism, whether human or of lower grade, the terms Lever-arm and Moment are applied. The Lever-arm of Power is the distance, measured on the lever, between the point of application of power and the point of application of the fulcrum, and the Lever-arm of Weight is the distance, measured along the lever, between the point of application of the weight and that of the

FIRST ORDER · P F W

P   F   W

SECOND ORDER · F W P

F   W   P

THIRD ORDER · F P W

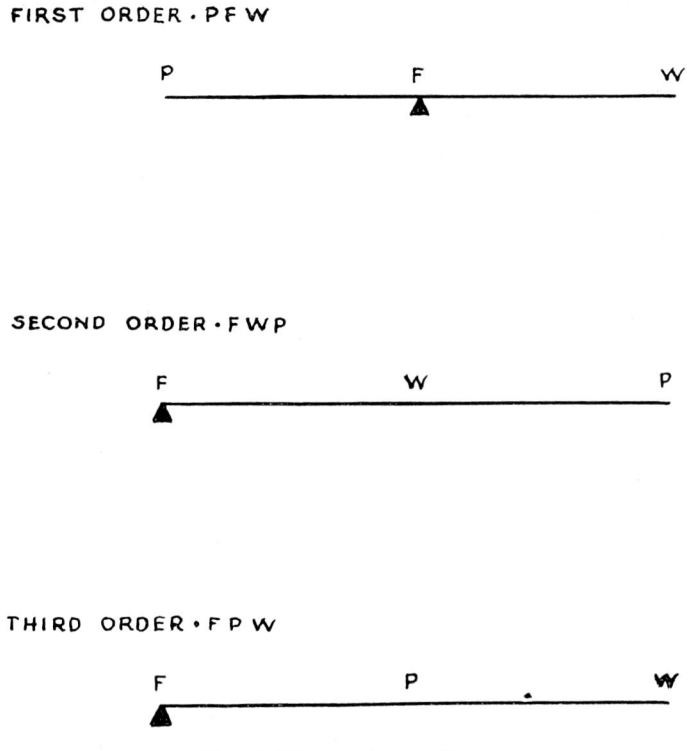

F   P   W

FIG. 1. Three orders of lever

fulcrum. In animal mechanism the point of application of power and the point of application of the fulcrum are easily obtainable, but the point of application of weight (or resistance) is difficult. In general terms, the point of application of weight is at the center of gravity of the moving body segment, and the Lever-arm of weight is the distance from the fulcrum to that center of gravity, if the segment is free moving. If there is definite resistance to the lever, the point of application of weight is the position of this resistance. (Note: Of course, if the segment has an added weight, as a heavy dumb-bell carried in the hand, when discussing, for example, the leverage action of the Biceps muscle on the forearm and hand, the center of gravity of the moving segment and the Lever-arm of weight would be shifted further away from the fulcrum than is the case without such addition). To calculate problems in Leverage, the physicist also measures the amount of the power and the amount of the weight, in pounds or kilograms. This measurement is also difficult, in animal mechanism, but we can get an approximate estimate of the power, at least, in the comparative bulk of the muscle or muscles used. The Moment of Power is the amount of power multiplied by the Lever-arm of power, and is stated in "Foot-pounds" or "Kilogram-meters"; and the Moment of Weight is the amount of the weight times the Lever-arm of weight and is measured as is the power, with similar statement of results.

The discussions in this book must necessarily omit this second element in leverage problems, except for an occasional reference to muscle bulk, but will discuss orders of leverage and Lever-arm efficiency somewhat fully. We here note that short arm-length of power tends to swiftness of motion of the moving segment and the necessity for greater bulk of the motive muscle or muscles; whereas, longer arm-length of power tends to greater efficiency of the action and requiring relatively less muscle bulk. This principle is well illustrated by the gearings of an automobile motor. High gear, with the smaller radius of its hidden gear-shift wheel, makes for high speeds in the car; low gear, having a larger gear-shift wheel, insures greater mechanical power to the car, as in hill climbing or with heavy trucks.

Chapter 1

FUNDAMENTAL STANDING POSITION

(Fig. 2, male, and Fig. 50a, p. 168, female)

Gymnastically described, in *Fundamental Standing Position*, the body is erect, supported by two feet, heels together, toes at an angle of ninety or sixty degrees, each to each, or with feet parallel and apart, knees fully extended, hips slightly forward, the plane of the inlet of the pelvis at fifty to sixty degrees with the horizon, chest elevated and somewhat forward, head and neck drawn back, face forward, shoulder blades adducted and hands at sides of thighs with palms facing inward.

The *line of gravity* of the body passes from the vertex of the head, through the transverse line joining the occipital condyles, in front of the tenth thoracic vertebra, through the line joining the centers of the hip joints, a trifle behind the line joining the centers of the knee joints and reaches the base of support just in front of the line joining the ankle joints.

The *base of support* is that area, on the floor, which is bounded by the outer margin of the right foot, the line drawn between the tips of the toes, passing laterally from the tip of the right great toe to the tip of the left great toe, to the tips of the toes of the left foot, the outer margin of the left foot and the line drawn from the rear of the heel of the left foot, to the starting point at the rear of the heel of the right foot (Fig. 3). (Note: It is obvious that the further apart the feet are held, the greater is the base of support; also that, if the heels are held together and the toes turned out, the greatest area of support will be attained with the feet at ninety degrees with each other. However, as the right angled position is recognized, on the one hand, as a position of much strain to the plantar arch of the foot and as the straight forward position, on the other hand, is that from which forward progression, as in marching, is most readily started, more recent Army regulations and gymnasium floor instructions direct a com-

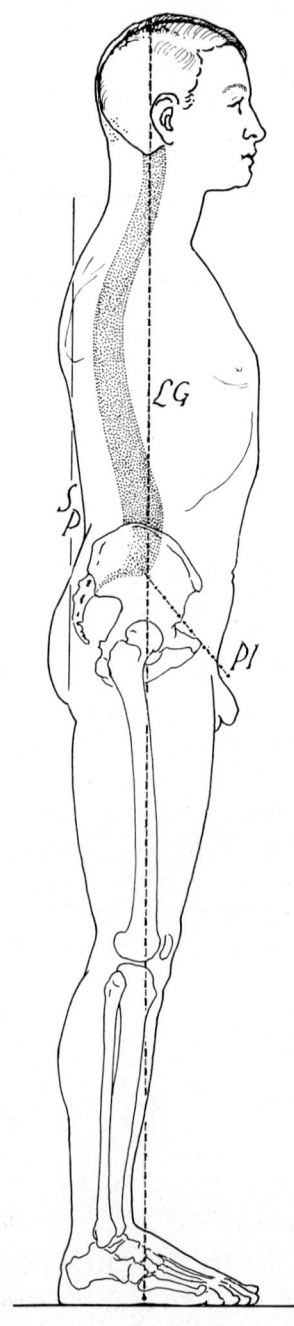

Fig. 2. Fundamental standing position, male, side view, $\frac{1}{10}$ life (see also Fig. 50, Normal, p. 168).

promise position at about thirty degrees. The base of support widens materially in all stride-standing positions; narrows in close-standing position and especially in half-standing position. The figure of the Running Mercury, in statuary, represents a remarkably small base of support.

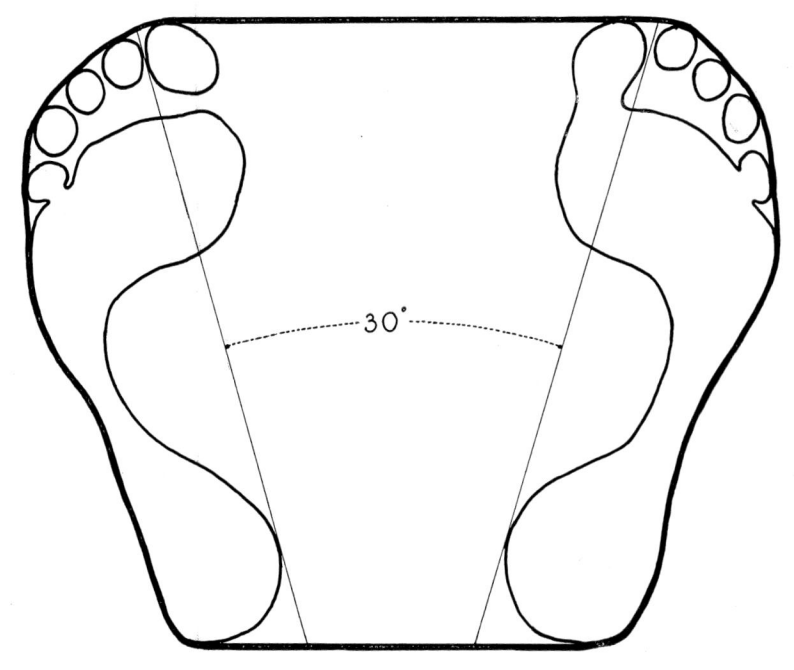

FIG. 3. Base of support, ⅓ life

THE FOOT

There are twenty-six *bones* in each foot, seven tarsus, five metatarsus and fourteen phalanges (Fig. 5b, p. 20). The shapes of these bones and the planes of their faces, where each joins its neighbor or neighbors, are so arranged as to favor the maintenance of the normal arches of the feet. Also careful study of the internal structure of the bones of the foot shows that the main supporting bones, especially the os calcis, the astragalus and the heads of the metatarsals, are so arranged, in the lines of their trabiculae, as to favor strength in stance. Three arches to the foot are described (Fig. 4). They are the long or plantar, the metatarsal

16   BODY POISE

or anterior transverse[1] and the mid-transverse. The plantar arch is based at the rear by the os calcis and at the front by the

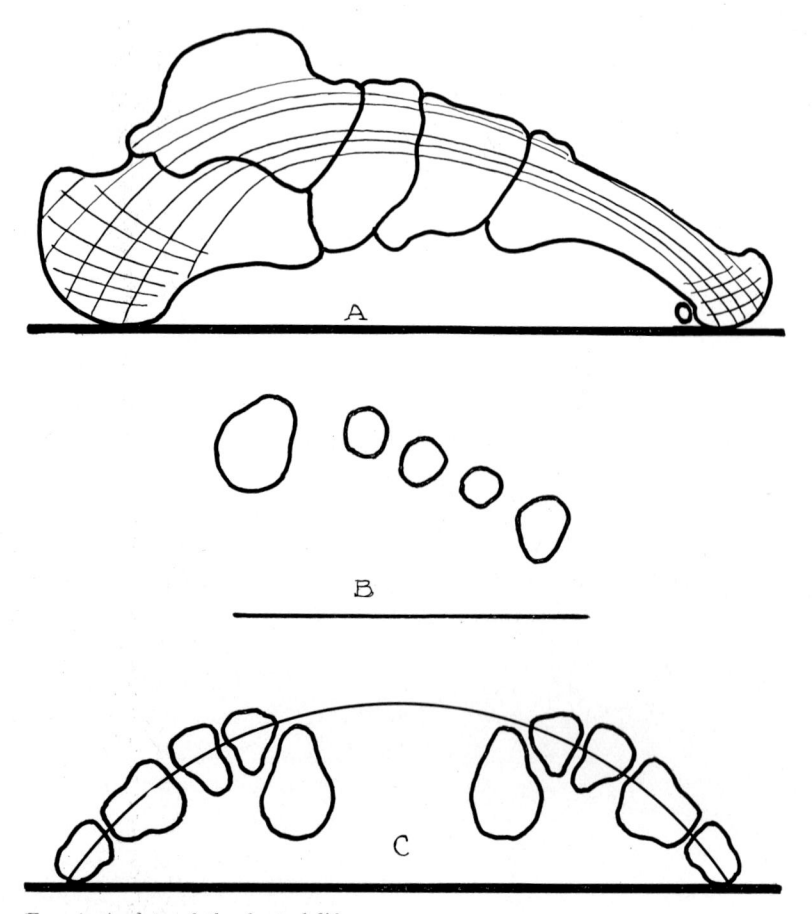

Fig. 4. Arches of the foot, ½ life.
A. Long or plantar.
B. Anterior or metatarsal (just back of metatarsal heads).
C. Midtarsal.

heads of the metatarsal bones, especially those of metatarsal I and metatarsal V; all of the bones of the tarsus and of the meta-

[1] Morton, D. (The Human Foot) maintains that, contrary to the usual teaching, there exists no metatarsal arch, and that the heads of the three middle metatarsal bones rest normally upon the ground, in the standing position.

tarsus contribute to this arch, whose highest or "key-stone" is the astragalus. Although there is no true anterior transverse arch at the metatarsal heads, the three mid metatarsal shafts are elevated. (See p. 138.) As a complete arch, the mid-transverse is only half present in each foot. It requires both feet to make the entire arch. So it may be defined, anatomically for each foot, as based on the outer border—the fifth metatarsal shaft, and the soft tissues beneath the cuboid and the anterior portion of the os calcis—, as rising upward inward beneath the proximal ends of the remaining metatarsal bones, the three cuneiform bones and the navicular and attaining its highest point (for that foot) beneath the astragalus and the sustentaculum tali of the os calcis. To complete the arch, which theoretically rises somewhat higher in the intermediate space between the feet, the downward descending structures of the other foot must be included. For purposes of description these three arches—two complete and one only half-complete in each foot—are named; but from the standpoint of treatment, as will be seen in Chapter 2, to consider the foot as anything but one complete structure leads to faulty care of mechanical disorders. (See p. 149 and p. 152).

Our feet are supplied with these wonderful arches not primarily for standing but for locomotion—walking, running, leaping and, in our arboreal ancestors, for tree climbing. Were the life of man a static one, simple broadened pedestals would be more efficient.

The facile arched formation of the foot is maintained by a) the shapes and inter-relationships of the bones themselves (see above); by b) ligaments; by c) muscles and tendons.

Of the intrinsic *ligaments* of the foot, we are most interested (in the discussion of human stance) in a) the long plantar ligament, extending from the under surface of the os calcis, forward to the cuboid bone and to the heads of metatarsals II, III, IV and V; b) the short plantar ligament, extending from the os calcis, forward inward to the cuboid. (These ligaments are strong and help materially in maintaining the long and the mid-transverse arches of the foot); c) the anterior transverse ligament, connecting the heads of the various metatarsal bones, each with its neighbor. (This ligamentous group aids in maintaining the anterior transverse arch of the foot).

Many other ligaments connecting neighboring bones, of the tarsus, of the metatarsus and of the phalanges and each with other, on the plantar, the dorsal and the lateral surfaces of the foot, add to the integrity of position of the bones of the foot, but do not play the important part in stance as do the three here specified.

MUSCLES AND MOVEMENTS OF THE FOOT (FIG. 9B, p. 28)

The movements of the foot are adduction or supination, abduction or pronation, flexion at the midtarsal and at the toe joints and extension at the midtarsal and at the toe joints.

a. *Adduction* or *supination*, with inward rotation of the forepart of the foot, takes place at the calcaneo-astragaloid, the calcaneo-cuboid and the naviculo-cuneiform joints. It is *caused* by Tibialis anticus, Tibialis posticus, Flexor longus hallucis and Flexor longus digitorum, which have their origin above the ankle joint; also the following shorter muscles, situated entirely in the sole of the foot—Flexor accessorius pedis, Lumbricales, Flexor brevis hallucis, Adductor hallucis, Flexor brevis minimi digiti, Interossei and Transversus pedis. Adduction is *limited* by interosseal ligaments on the outer side of the tarsus and of the ankle joint.

b. *Abduction* or *pronation*, with outward rotation of the forepart of the foot, takes place at the calcaneo-astragaloid, the calcaneo-cuboid and the naviculo-cuneiform joints. It is *caused* by Peroneus longus, Peroneus brevis and Peroneus tertius. Abduction is *limited* by the plantar ligaments and by interosseous ligaments on the inner side of the tarsus and of the ankle joint.

c. *Flexion* at the mid-tarsal joints and flexion of the toes is *caused* by Tibialis anticus, Flexor longus digitorum and Flexor longus hallucis, which have their origin above the ankle joint; also the following which are located entirely in the sole of the foot —Abductor hallucis, Flexor brevis digitorum, Adductor minimi digiti, Flexor accessoris pedis, Lumbricales, Flexor brevis hallucis, Adductor hallucis, Flexor brevis minimi digiti and Interossei.

Flexion is *limited* by the interosseous ligaments on the entire dorsum of the foot, including those of the toes.

d. *Extension* at the mid-tarsal joints and extension of the toes is *caused* by Extensor longus digitorum, Extensor proprius hallucis, Peroneus longus, Peroneus brevis and Peroneus tertius. Extension is *limited* by the plantar and interosseous ligaments on the plantar side of the foot, including those of the toes.

It will be noted that the muscles used for adduction, or supination and inward rotation of the forepart of the foot and for flexion of the toes, are more numerous than those used for abduction, or pronation and outward rotation of the fore part of the foot with extension of the toes. This is, of course, necessary to counteract the effect of weight-bearing on the foot in Fundamental Standing Position.

To recapitulate, it is to be noted that, were the human foot function one of standing only, a pair of broad, bony pedestals would have been useful; that to supply man's need for facile bases for locomotion—walking, running, leaping, dancing, skating—our feet are supplied each with three arches, but that, in standing, superimposed weight puts much strain upon these arches; that to meet this strain we are supplied with nicely shaped foot bones, with ligaments appropriately placed and with muscles which not only aid the supportive function of the bones and ligaments but which activate the feet in the many and varied uses of locomotion. Also, it is opportune, at this place, to anticipate the discussion of the mechanical disorders of the feet by stating that the line of defense against sagging due to weight-bearing must be considered in the reverse order of that which has been here named; that is, first the muscles and tendons stretch and lengthen and weaken, then the ligaments stretch and lengthen and weaken and finally faulty positions, long held, cause structural changes in the shape of and the facial approximations of the bones each to the other.

THE LEG

The leg is composed of two bones (Fig. 5a, p. 20)—a larger tibia on the inner side and a smaller fibula on the outer side. Both are long bones. The tibiae receive all of the weight of struc-

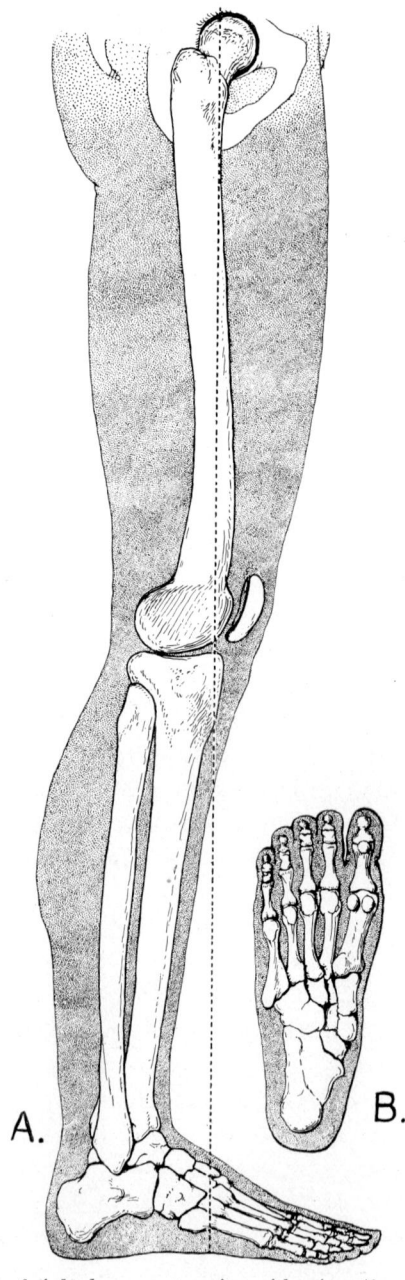

Fig. 5. A. Bones of right lower extremity, side view (female), $\frac{1}{5}$ life.
B. Bones of the right foot, plantar view, $\frac{1}{5}$ life.

tures above them and transmit practically all of the superimposed weight to the astragali below. The *Tibia* is broadened above and has an upper surface nearly horizontal, in the standing position, but with two shallow cup-like depressions to receive and to guide the motions of the two condyles of the femur. The shaft of the tibia narrows from above downward, although in its entire length it is much more massive than is the fibula. At the lower end, the tibia again becomes more bulky and is so modified as to afford about two-thirds of the articular surface for joining with the astragalus. Of this lower articular surface, one portion is practically horizontal for weight-bearing on the astragalus; the other portion is perpendicular (in the standing position) to form one-half of the mortising of the astragalus. The mechanical purpose of the tibia is weight-bearing and leverage in locomotion and other body activities. The *Fibula* is a long bone placed at the outer side and (above) a trifle to the front of the tibia. It is much less massive than its mate and is attached to it by ligaments above and below and by a strong membranous septum along the entire length of both bones. It has no articulation with the femur above, although an enlarged head gives attachment to certain ligaments and muscles. The shaft is irregularly rod-like and the lower end is modified to form the external malleolus, with its perpendicular articular surface, to complete the mortising of the astragalus. There is no motion between the tibia and fibula and such ligaments as exist between them are for the sole purpose of holding the fibula firmly to the tibia. The two bones are thus to be considered as a single segment in body mechanism. As such, they afford attachment to ligaments of the ankle joint below and of the knee joint above. With the strong septum between the bones, the tibia and fibula give *origin* to the Flexor longus digitorum, Flexor longus hallucis, Tibialis anticus, Soleus, Tibialis posticus, Peroneus longus, Peroneus brevis, Peroneus tertius, Extensor longus digitorum and Extensor proprius hallucis, acting over the ankle joint, on the foot and on the toes below. Above, this mechanical segment affords *insertion* to Biceps femoris, Quadriceps femoris (by means of ligamentum patellae), Gracilis, Sartorius, Semitendinosis, Semimembranosis and Popliteus, acting from above, over the knee joint, upon the leg.

THE ANKLE JOINT

The ankle joint is a ginglymus or hinge joint, allowing simple flexion and extension, which (for this joint only) is called dorsi-flexion and plantar-flexion. However, with increasing plantar-flexion, an increasing amount of adduction and abduction of the foot is attainable. The bony mechanism of the ankle joint consists of an encompassing mortise, formed by the lower end of the tibia and of the fibula, into which the astragalus, fits allowing free motion forward and backward.

The *ligaments* of the ankle joint are a) anterior, extending from the internal malleolus of tibia, from the lower anterior border of tibia and from the external malleolus above to the rough upper surface of the neck of the astragalus, and limiting the movement of plantar flexion of the foot; b) posterior, extending from the internal malleolus and rear lower border of tibia above to posterior upper surface of astragalus below, and limiting the movement of dorsi-flexion of the foot; c) internal lateral ligament (deltoid), extending from tip and the entire anterior margin of the internal malleolus above to the inner side of the body of the astragalus and to the tip of the sustentaculum tali, limiting the abduction, pronation and outward rotation of the fore part of the foot (noted above), and reinforcing the bony malleoli in resisting lateral strains at the ankle joint; d) external lateral ligament, three distinct bands, extending from the anterior and the posterior borders of the external malleolus above to the upper outer front, the upper outer rear of the astragalus and to the outer side of the os calcis. They limit adduction, supination and inward rotation of the fore part of the foot (noted above), and reinforce the bony malleoli in resisting lateral stresses at the ankle joint. (It is to be noted that some so-called sprains of the ankle joint include actual separation of the tip of the internal or external malleolus, the small fragment of bone being torn from the tibia or fibula, respectively, rather than rupturing the lateral ligament).

Movements and *Muscles* of the Ankle Joint (Fig. 9a, p. 28): a) *Plantar flexion* of the foot is *caused* by Gastracnemius, Soleus, Flexor longus digitorum, Flexor longus hallucis and Tibialis posticus; is *limited* by approximation of the rear of the astragalus and lower rear of tibia, by the anterior fibres of the internal and external lateral ligaments and by the anterior ligament and is *opposed* by the Dorsi-flexor muscles of the foot; b) *Dorsi-flexion* of the foot is *caused* by Tibialis anticus, Extensor longus digitorum, Extensor longus hallucis, Peroneus longus, Peroneus brevis and Peroneus tertius; is *limited* by the neck of the astragalus abutting on the lower front of the tibia and by tension of most of the fibres of the internal lateral, the middle and rear fibres of the external lateral and the posterior ligaments of the ankle joint, and is *opposed* by the plantar muscles of the foot, especially the calf muscles.

KINESIOLOGY OF THE ANKLE JOINT

Considering the movements and limitations of the ankle joint from the view point of Fundamental Standing Position, we note that these may be

a. forward swaying of the entire superstructure (Fig. 6), which is initiated by the dorsi-flexors of the foot and carried on by the force of gravity, and is limited by plantar-flexing muscles (the calf muscles), by the posteriorly placed ankle joint ligaments and by the approximation of the anterior portions of the bony structures immediately involved in the ankle joint. As the line of gravity passes definitely in front of the ankle joints, in this modification of Fundamental Standing Position, there is a continuous mild tonic contraction of the calf muscles to maintain the position.

b. backward swaying at the ankle joint is initiated by the plantar flexors and continued by the force of gravity on the superstructure, and is limited by dorsi flexor muscles, by the anteriorly placed ankle joint ligaments and by the abutting of bony structure involved in the ankle joint and to the rear of the same.

LEVERAGE INVOLVING THE ANKLE JOINT

1. From the view point of Fundamental Standing Position, the gymnastic exercise of heel raising uses the foot as a lever of the second order—FWP (see Fig. 7): the foot, held rigid by its intrinsic muscles and ligaments, is the lever; the fulcrum is at the base of the toes, the weight is the entire body structure above and applied at the ankle joint, and the power is the calf muscle applied,

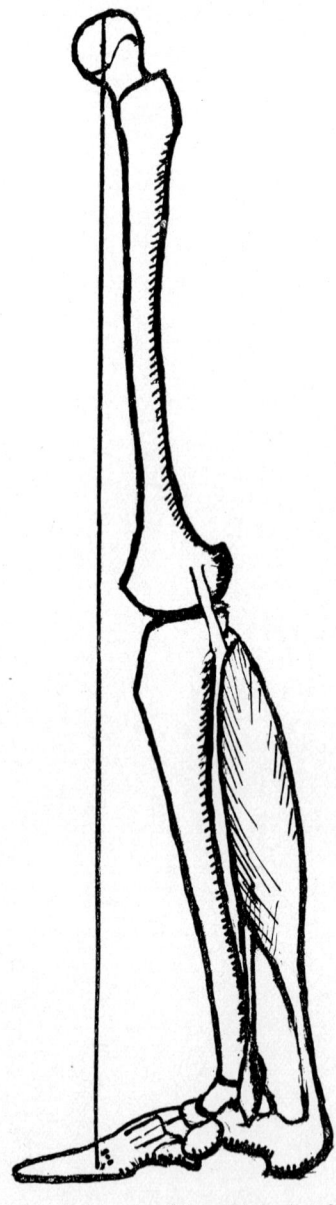

FIG. 6. Standing, forward swaying at ankles, $\frac{1}{5}$ life (modified from Mollier)

through the Tendo Achillis, at the rear of the os calcis. As the power-arm distance is but little greater than the weight-arm distance, the leverage efficiency is not ideal. Considerable bulk

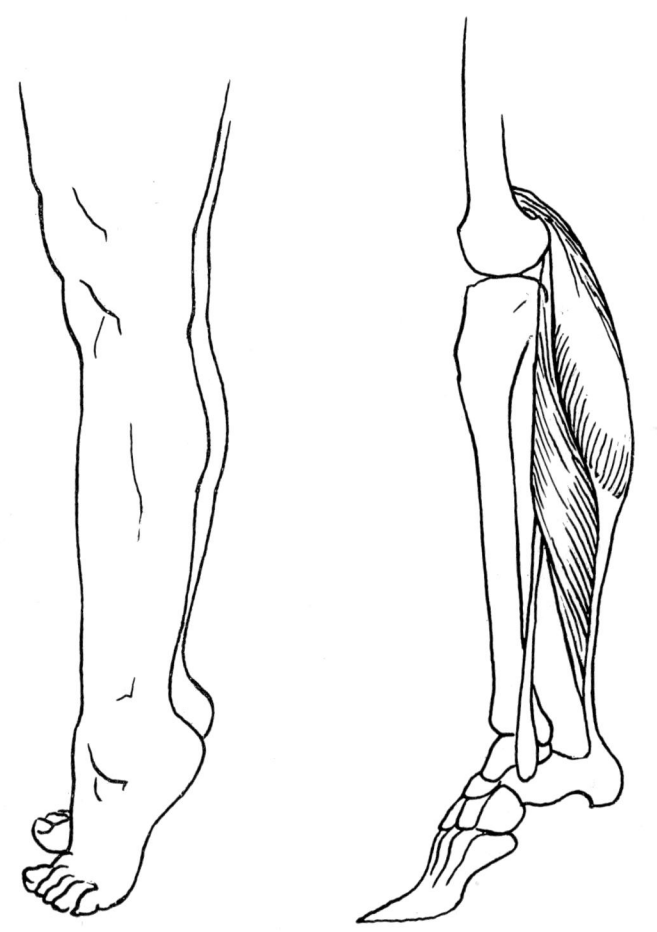

Fig. 7. Calf muscles in tip-toe standing, $\frac{1}{5}$ life, foot lever of second order (modified from Mollier).

(and therefore power) in the calf muscles here meets this deficiency.

(Note: In the slight compensatory swaying of the superstructure—plantar-flexion—at the ankle joint, to keep the frame (above that joint) from toppling forward—we might consider a

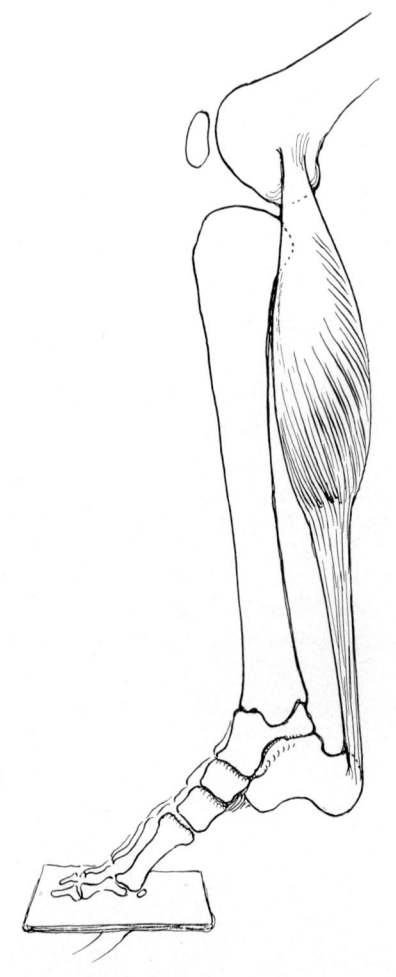

FIG. 8. Calf muscles in auto foot-breaking, $\frac{1}{5}$ life, foot lever of first order

compensatory leverage of the body, with the ankle joint as a fulcrum. What actually takes place, is a maintenance of the superstructure, in its normal line of gravity position, by a generally added tonicity of the plantar flexors in concentric contraction and the dorsi flexors in eccentric contraction).

2. Were we here considering a body position of sitting at the controls of an automobile (Fig. 8), for instance, with the balls of the foot on the brake pedal, foot pressure upon the pedal makes use of the foot as a lever of the first order: the ankle joint is the fulcrum, the calf muscles, applied at the rear of the heel, is the power and the resistance of the brake pedal, applied at the balls of the foot, is the weight element—PFW.

THE THIGH

The two thighs constitute the next segment in weight-bearing, in Fundamental Standing Position. All of the body-weight above them is borne upon the heads or upper ends of the thigh bones; and this weight plus their own weight rests upon the tibiae below.

The *femur* (Fig. 5a, p. 20) extends from the hip joint to the knee joint. It is the longest bone of the body. It has a shaft; a head, a neck and a greater and a lesser trochanter at the upper end, and two condyles and an inter condylar notch at the lower end. It joins with the tibia at the lower end and with the hip bone at the upper end. The femur gives attachment to certain ligaments at the upper end—capsular, transverse and ligamentum teres. Below, the femur gives attachment to ligaments regulating the knee joint. The thigh bone affords *origin* to Gastrocnemunius and Plantaris, acting over the knee joint on the foot; to Biceps femoris, the greater part of the Quadriceps femoris (Vastus externus, Vastus internus and Crureus) and the Popliteus, acting on the knee joint by their insertions in the leg. The femur gives *insertion* to Pyriformis, Obturator internus and the Gemelli, Quadratus femoris, Gluteus maximus, Gluteus medius, Gluteus minimus, Adductor longus, Adductor brevis, Adductor magnus and Obturator externus, acting from the pelvis on the femur; and to the Ilio-psoas muscle acting from the lumbar spine, over the low inter-spinal joints and on to the femur (Fig. 9).

The mechanical function of the thighs is weight-bearing and leverage in motion and locomotion.

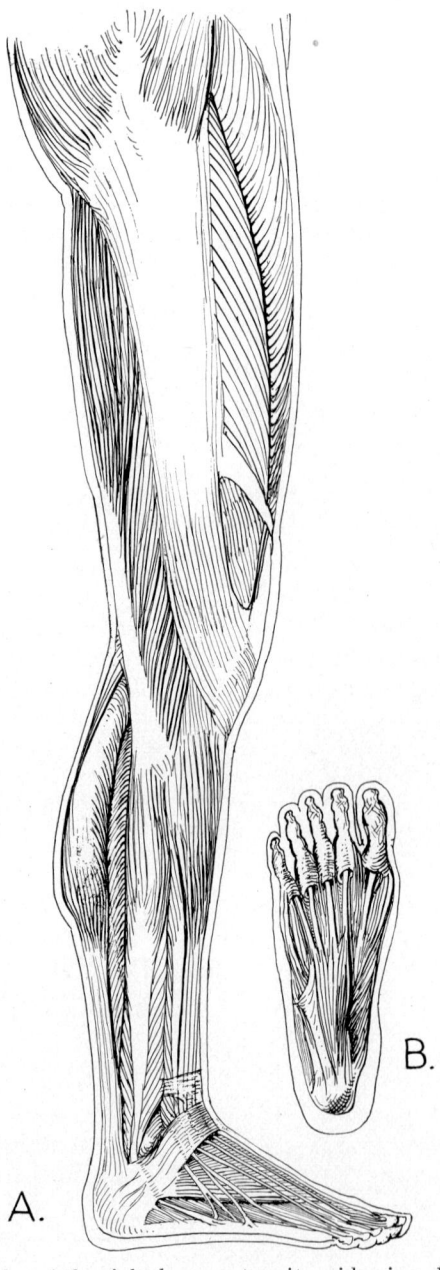

FIG. 9. A. Muscles of the right lower extremity, side view, $\frac{1}{5}$ life.
B. Muscles of the sole of the right foot, $\frac{1}{5}$ life.

THE KNEE JOINT

In the study of Fundamental Standing Position, the next movable part is the knee joint. This, too, is a ginglymus or hinge joint, allowing simple flexion and extension only to the straight limb position for most of its range of motion. However, with increasing flexion, there may be a certain amount of rotation inward, and in return to extension, a derotation (in each case, the tibia pivoting on the inner and gliding on the outer condyle of the femur). The bony mechanism of the knee joint consists of the shallow cupped upper surface of the tibia (two shallow cups placed side by side) and the two condyles of the femur. The knee joint is the largest in scope and in articular area in the body. It is subject to much stress and strain and injury in bodily activities. It is poorly supplied with natural bony protection to these stresses. To meet this it is well supplied with ligamentous and muscular safeguards.

The *patella* is a small bone situated in front of the knee joint. It affords anatomical insertion above to the strong Quadriceps extensor muscle and below to the strong ligamentum patellae, which carries the physiological insertion of the Quadriceps extensor muscle to the upper front of the tibia, thereby extending the action of the Quadriceps extensor to the tibia. To the sides of the patella also are attached fibrous expansions of the Quadriceps muscle. The patella, by virtue of its actual bulk and because of its muscular and ligamentous attachments, is a useful protector of the knee.

The *ligament* proper of the knee is a capsular ligament, extending from the lower end of the femur, all around the latter's circumference and just beyond the articular surfaces of the condyles, to the entire circumference of the upper margins of the tibia just below the upper articular surface of tibia. This capsular ligament is reinforced by stronger augmenting bands at front, at rear and at sides of the knee joint. Those in the rear definitely aid in limiting extension of the leg at straight limb position. Within the knee joint are two semi-lunar cartilages (menisci), fixed to the right and the left outer rims of the upper articular surface of the tibia, and two crucial ligaments. The semi-lunar cartilages act as "fences", giving extra protection from side slipping of the condyles of the femur, in ordinary and in extra-ordinary activities. They definitely deepen the articular receiving cups of the tibia. The crucial ligaments are anterior and posterior. The anterior crucial ligament is strong. It is attached below to a bony spine, at about the middle of the top of the tibia, extends upward, backward and outward to be fixed to the inner surface of the external condyle of the femur. It specifically limits extension of the leg beyond the straight-limb position and to a lesser degree outward rotation of the leg in the partly flexed position of the leg. The posterior crucial ligament is stronger, shorter and less oblique than the anterior. It is attached below to the mid rear of the upper surface of the tibia, extends upward inward and forward, to be attached to the inner surface of the internal condyle of the femur. It, too, limits extension and inward rotation of the partly flexed leg. These crucial ligaments form the principal means of preventing leg extension beyond the straight limb position.

Movements and *Muscles* of the *Knee Joint:* (Figs. 10 and 11) a) *Flexion* of the leg is *caused* by the Gastrocnemius, Popliteus, Sartorius, Gracilis, Biceps femoris, Semitendinosis and Semimembranosis; and is *limited* by the approximation of the calf upon the rear of the thigh. b) *Extension* of the leg is *caused* by one powerful muscle—the Quadriceps extensor femoris; and is *limited*, at the straight-limb position, by tightening of the two crucial ligaments and the posterior thickened capsular ligament.

Some outward and some inward rotation at the knee joint is possible, when the leg is partially flexed. c) *Inward rotation*, in knee flexion, is *caused* by Popliteus, Sartorius, Gracilis, Semitendinosis and Semimembranosis; and is *limited* by the bony approximations and by the capsular and crucial ligaments. d) *Outward rotation* or derotation from knee flexion is *caused* by Sartorius and Biceps femoris; and is *limited* by the bony approximations and the capsular and crucial ligaments.

From the standpoint of Fundamental Standing Position, the movements of the knee joint are comparatively simple. Normally, the knees are "locked" in extension—that is, extension cannot take place beyond the straight-limb position, which is that of standing. However, to maintain this, a certain amount of tonic contraction of the Quadriceps femoris muscle must exist. If forward "give" (flexion) at the knee occurs, it is due to lessening of this tonic contraction of the Quadriceps.

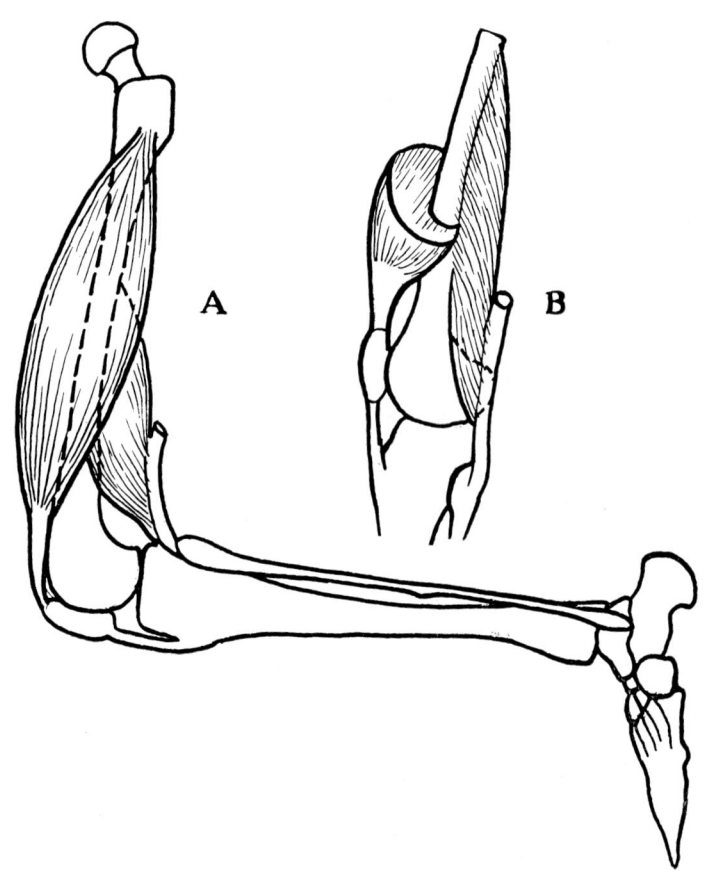

Fig. 10. Knee flexor and extensor muscles, about ⅕ life (modified from Mollier).
 A. In flexed knee position.
 B. In extended knee position.

Certain reciprocal actions of the muscles acting on both hip and knee joints will be discussed after the anatomical discussion of the hip joints. (See p. 42.)

KINESIOLOGY OF THE KNEE JOINT

In Fundamental Standing Position, less variation from the standard bone relationships is found in the double knee-joint mechanism than in other major articulations. The ligamentous structure of the joints prevents over extension. Greater or less flexion "give" of both knees is possible, but rarely present in vigorous health, although in "Gorilla" slouch (Fig. 50c, p. 168) it is a factor in general body relaxation. Knee "give" must include extra strain on the knee extensor muscles (Quadriceps), and is thus unconsciously avoided. However, a not uncommon variation is a flexion give of one knee, the other limb taking the brunt of support of the human structure—the so-called "debutante slouch". This unilateral give affects posture to a large degree. On the relaxed side besides flexion of the knee, there is plantar flexion of the foot, flexion of the hip and downward lateral tilt of the pelvis (to the same side). This relaxation, with unilateral flexion, causes a slight swaying of the lower trunk to the opposite side and, to keep the trunk within the line of gravity of the body, the upper trunk bends laterally slightly to the opposite side. This trunk weight shift and upper trunk bending, causes lateral deviation of the lumbar spine (to the same side). On the supporting limb, there is abduction at the ankle joint, no change at the knee joint except that its Quadriceps must be in greater tension, because of the extra work it has to do, and adduction at the hip.

This unilateral knee-bend standing is usually transitory. It may be taken to ease general limb strain or foot strain. It is rather apt to involve alternate limbs. Like Zekle, in Lowell's "The Courtin' ", it may also express embarrassment:

> "He stood a spell on one foot fust
> Then stood a spell on t'other,
> An' on which foot he felt the wust
> He couldn't ha' told ye nuther."

But if this knee give becomes habitual on one side more than another, it may be one of the potent causes of functional and later structural curvature of the spine (pp. 171 and 199).

LEVERAGE AT THE KNEE JOINT

In Fundamental Standing Position, there is no leverage problem at the back of the knee. The knee is already fully extended and the intrinsic ligaments prevent further forward bending of

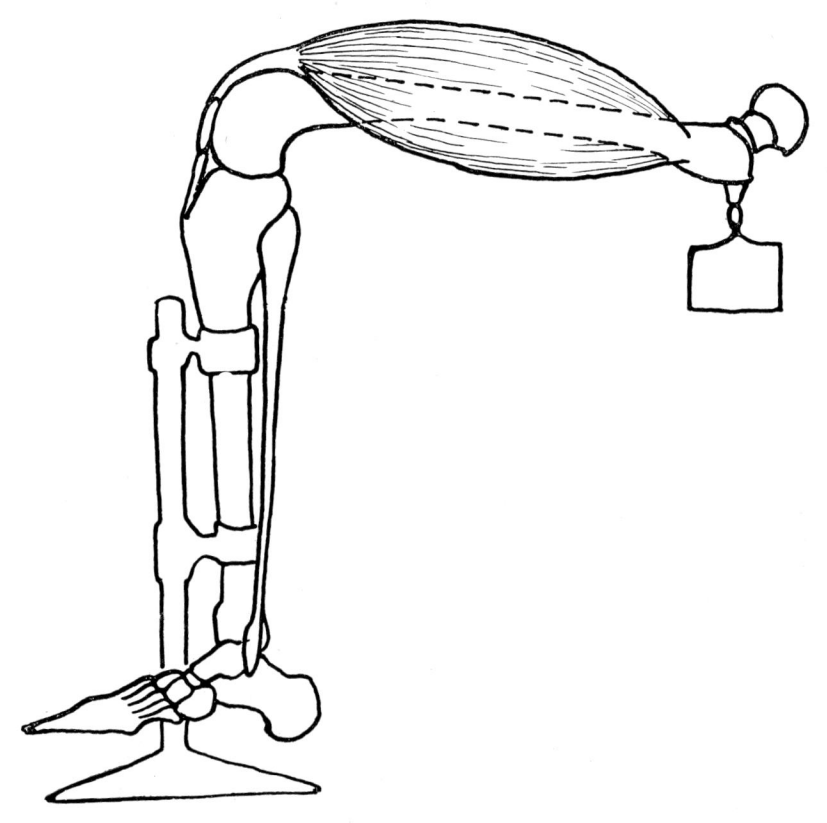

Fig. 11. Eccentric action of Quadriceps femoris muscle, in unusual flexed knee position, about ⅕ life (modified from Mollier).

the superstructure. At the front of the knee, leverage may be discussed only on the supposition that the feet and legs are so fixed as to make backward bending at the knees possible (Fig. 11, p. 33). In this rare circumstance, the lever is composed of the entire superstructure, fixed in its many segments by many muscles and ligaments to become one moving segment; the fulcrum is at the knee joints, the power is the Quadriceps muscle group, acting from insertion to origin onto the front of the thigh, and the weight is the center of gravity of the entire superstructure—trunk, head and neck and upper extremities. This point of application of the weight is considerably further along the lever than is that of the power. The lever is of the third order, FPW. It is obvious that if the legs and feet can be so firmly held, by some outside force, as to allow much backward bending of the superstructure, the moment of the weight is greatly increased; and if it were possible so to bend the trunk backward until the knees were at right angled flexion, with the superstructure pendant, as it were, at horizontal, the strain on the eccentrically contracted muscles would be almost prohibitive. With the long lever-arm attachment of power, this leverage is a good example of force of action rather than of speed.

Although the foregoing bodily activity is a rare, though possible one, in ordinary life, it is very common as an alternate knee function in walking, in running and going up stairs. In walking (see Fig. 89, p. 244) for instance, immediately following the placement of the forward foot upon the ground, the knee of the same side straightens (extends) and for that side the leverage problem is the same as that discussed above for both sides. It is a lever of the third order, and the muscles (Quadriceps group) act concentrically from insertion to origin. As some of the body weight is carried forward and the opposite limb completes its forward excursion to its new foot placement, the same leverage becomes operative for the alternate side. In running and in stair climbing, the same alternate principles of knee-joint leverage are invoked; but in stair and hill climbing, the force exerted by the Quadriceps must be greater because each set of muscles must not only carry the body weight forward but must lift it as well.

In sitting, with leg extension as an exercise, the principle of leverage is reversed. Here the lever is the leg and foot; the fulcrum is at the condyles of the femur, the power is applied through the patella to the upper front of the tibia and the weight is concentrated or applied to the center of gravity of the moving part. It is a lever of the third order, FPW. Here the weight is to be moved and is much less than in the previous conditions, and much less force is required to accomplish it. This is a good example of potential speed of motion, mechanically attainable because of short power-arm.

To this point, we have described the body segments and the joints as two-fold, that is, with a bifid base. Now we deal with single segments—the pelvis, the twenty-four bones of the vertebra and the head—and then with bilaterally symmetrical segments—the chest and those of the upper extremities.

THE PELVIS (FIG. 14, p. 46, AND FIG. 15, p. 60)

The *pelvis* consists of three bones, the right and the left ossa innominata and the sacrum. (The small coccyx is a fourth bone but is of but little note in this discussion). These three major bones are normally so firmly united with each other as to consider them as one segment. The pelvis is of irregular shape. It is much broader above than below. It carries the weight of all the body above it, at its upper or lumbo-sacral joint. At its sides, the two deep acetabular cavities transmit the weight of the entire trunk, head and arms to the thighs at the hip joints. The massive sacrum, which supports so much weight from above, is situated at the rear. On either side of the sacrum are broad flanges of bone which have the double purpose of supporting much of the abdominal and pelvic contents and of giving attachment to important muscles extending upward and downward. These flanges are part of the *ossa innominata*, which bones circle around forward and downward to be firmly united to each other in front by ligaments and cartilage, at the symphysis pubis. Within this bony encirclement is a complete opening, which is divided, for convenience of description, into the pelvic inlet and the pelvic outlet. These forward-encircling bones afford protection to the pelvic contents, added support to the abdominal contents, a not inconsiderable leverage-arm for trunk and for thigh flexions and give attachment to muscles acting above and acting below. The opening within these encircling bony arms contains the pelvic contents, the rectum, the bladder, the uterus or the prostate, and through it passes in women the newborn baby. The *sacrum*, situated at the rear of the pelvis, is also classed as one of the vertebrae. It is concave forward and convex backward. It has a body and articular processes above for joint union with lumbar vertebra V and through its downward curved length there is a canal for the passage of the spinal cord and blood vessels.

The *pelvis as a whole* affords attachment first to strong front and rear sacro-iliac joint and symphysis pubic ligaments, which firmly unite these joints, and to cushioning fibro-cartilages which ease the shocks of bodily stresses and strains. Certain ligaments —the great sacro-sciatic and the lesser sacro-sciatic—pass between the sacrum and the ossa innominata, on each side, to afford protection for the passage of important vessels and nerves and smaller muscles; and Poupart's ligament, on either side at the upper front of each os innominatum, for the passage beneath it of important vessels and nerves and of the Psoas magnus muscle and for the attachment of some of the abdominal muscles. In addition, certain ligaments connect the pelvis with the thigh bone on either side. They are the capsular, the transverse, the ligamentum teres and the cotyloid cartilages. Attached to the upper portion of the sacrum and to the nearby crests of the ossa innominata are the intervertebral fibro-cartilage, anterior common, posterior common, capsular, ligamentum subflava, supra-spinous, inter-spinous and ilio-lumbar ligaments. The pelvis affords *origin* to Pyriformis, Iliacus, Gluteus maximus, Gluteus medius, Gluteus minimus, Tensor vaginae femoris, Sartorius, Rectus femoris, Obturator externus, the Gemelli, Pectineus, Adductor longus, Adductor brevis, Adductor magnus, Gracilis, Biceps femoris, Semi-membranosus, Semi-tendonosus, Quadratus femoris and Obturator internus muscles, acting from the pelvis on the thigh bone or on the thigh bone and leg. The pelvis also affords *origin* to Erector spinae, Latissimus dorsi, Mutifidus spinae, Quadratus lumborum, Rectus femoris, Obliquus internus, Transversalis and Obturator internus muscles, acting from below upward on the spine or the spine and thorax or the spine, thorax and upper extremity. The pelvis affords insertion to practically no muscles.

In anatomical study of muscles, the word "origin" refers to that bone or other attachment which is the more fixed and the word "insertion" to that attachment which is the more movable in ordinary body activities. Thus we see that the pelvis, while affording origin to many muscles, acting both upward and downward, gives insertion to no muscles of importance to the human general motive or locomotive mechanism. (A few muscles, connected with the functions of defecation and parturition, have both origin and insertion on the pelvis and the coccyx). Viewed from several angles, this is what we would expect. For instance, the infant, lying in its crib, will first move arms and legs. Then there will be an attempt to lift the head. In either case, the pelvis and the trunk are comparatively immovable. Later there will be trunk twistings and gradual trunk liftings toward sitting with the pelvis still comparatively immovable. At first, in the case of the infant, the weak undeveloped muscles are called upon to lift the lesser weights, then greater body loads; but always the work is from center outward. Again, we might consider this eccentric action of muscles in adult occupations. Many of them involve long-time sitting but with more or less activity of head, of arms, even of legs and of trunk, the pelvis being the least movable of all. Striking examples of this arrangement of origin and insertion of muscles and their progressive extension from the pelvis outward, are found in the bicycle rider who, with his pelvis comparatively immovable upon the seat of the bicycle, works his limbs from pelvis to thighs, from thighs to legs and from legs to feet, and his trunk and arms, less vigorously, from most fixed points distally, in steering the bicycle. Another example is found in the oarsman, who sits upon a comparatively fixed pelvis and uses his trunk and arm and leg muscles vigorously, in propelling the boat. Probably the most scientific approach to this fascinating study, of the outward action of muscles, is to be found in tracing the principle through evolutionary developments in the animal kingdom—the simple trunk twistings of worms; the fishes, with uncomplicated fin and tail activities; the centipedes, with simple multiple addition of limbs; the belly-crawling salamanders and newts, which have more complex limbs and much activity of head and neck, and finally the birds, the mammals and man, with ever increasing variety and efficiency of motive and locomotive structure. Through all of these, biologists and comparative anatomists trace the same principle of relative fixation of control muscles and increasing movability of more and more distally applied muscles. From the point of view of this book, the principle is important because much of the disorganization and disabilities of posture are found in the central body mechanism; and correction, especially gymnastic correction, should be directed primarily to this region, and only secondarily to the body appendages.

Having described the anatomy, the kinesiology and leverage of the limbs—feet, ankles, legs, knee joints and thighs—and the anatomy of the pelvis in Fundamental Standing Position, the next movable mechanism is that of the hip joints.

THE HIP JOINT

The *hip joint* is an enarthrodial or ball-and-socket joint. The bony mechanism of the hip joint includes the femur and the pelvis. The head of the femur (a round ball about the size of a golf ball) is placed at the inner end of the neck of the femur; the neck being bent away from the shaft of the femur at somewhat more than a right angle. The os innominatum has a deep rounded cavity—the acetabulum—situated a little below the middle of its outer surface, into which nicely fits the head of the femur. This ball and socket mechanism allows free flexion, fairly free extension, adduction, abduction, inward rotation, outward rotation and circumduction of the femur. Unlike the knee joint, the hip is well supplied with bony protection; and has additional ligamentous and muscular support.

The *Ligaments of the hip joint* are the capsular, the transverse, the ligamentum teres and the cotyloid cartilage. The capsular ligament extends from just beyond the entire rim of the acetabulum, on the pelvis, outward downward to be attached below to the femur, beyond its neck (in front) and on its neck in the rear. The capsular ligament has three strong reinforcing bands, the ilio-femoral in front, the pectineo-femoral on the inner and lower sides of the joint and the ischio-femoral at the rear, to give added support and restraining motion to the joint where these needs are to be met. The cotyloid cartilage extends around the rim of the acetabulum and thus deepens its cavity. The transverse ligament augments the use of the cotyloid cartilage. The ligamentum teres extends from the bottom of the acetabular cavity to a small depression in the middle of the ball-like head of the femur.

Movements and *Muscles* of the *Hip Joint:* a) *Flexion* is *caused* by Adductor longus, Adductor brevis, Quadriceps femoris (Rectus only), Sartorius muscles, acting from the pelvis on to the leg; also, Iliopsoas muscle, acting from the spine and the pelvis directly on to the thigh. Flexion is *limited* by the approximation of the thigh on the abdomen and indirectly by the posterior ligaments of the hip joint. b) *Extension* is *caused* by Biceps femoris, Semitendonosus, Semimembranosus, Gluteus maximus, Gluteus medius muscles, acting from the pelvis directly on the thigh. Extension is *limited* by the strong ilio-femoral ligament of the hip joint. Extension in the hip is normally not beyond about thirty degrees of so-called "over extension". c) *Adduction* is *caused* by Adductor longus, Adductor brevis, Adductor magnus, Obturator internus and Gemelli, acting directly from the pelvis to the thigh bone, and Gracilis muscle, acting from the hip joint to the upper end of the leg; adduction is *limited* by the capsular ligament and the ligamentum teres of the hip joint and by the approximation of the thigh against the abducted opposite thigh. Adduction is normally about thirty degrees. d) *Abduction* is *caused* by Gluteus maximus, Gluteus medius, Gluteus minimus muscles, acting from the pelvis directly on the thigh bone; and Sartorius, Tensor fasciae latae, acting from the pelvis on the upper end of the leg. Abduction is *limited* by the approximation of the neck of the femur upon the upper lip of the acetabulum and by the capsular ligament and the ligamentum teres. Normally it is about forty-five degrees. e) *Inward rotation* is *caused* by Gluteus medius and Gluteus minimus muscles, acting from the pelvis directly on the femur, and the Sartorius, Gracilis, Semitendonosus, Semimembranosus, Tensor fasciae latae muscles, acting from the pelvis on the leg bones. Inward rotation is *limited* by the approximation of the neck of the femur with the front lip of the acetabulum and by the capsular ligament and the ligamentum teres. Normally about forty-five degrees are possible. f) *Outward rotation* is *caused* by Pyriformis, Obturator internus and Gemelli, Quadriceps femoris, Gluteus maximus, Adductor longus, Adductor brevis, Adductor magnus and Obturator externus muscles, acting from the pelvis directly on the femur, and Sartorius, Biceps femoris, acting from the pelvis on the leg bones. Outward rotation is *limited* by the approximation of the neck of the femur with the rear lip of the acetabulum and by the capsular ligament and ligamentum teres. Normally about sixty degrees are possible. g) *Circumduction* is that movement at the hip joint in which the limb describes a cone with the apex at the hip joint and the base at the sole of the foot. It is a combination of Flexion, Abduction, Extension and Adduction but without Rotations; and is *caused* by sequential combinations of the muscles described for each of these three motions and *limited* by the component factors of limitation.

Leverage at the Hip Joints

In Fundamental Standing Position, the lever is the entire superstructure above the hip joints. In forward trunk bending, the fulcrum is at the hip joints, the power is applied at the rear of the pelvis; the weight is applied at the center of gravity of the superstructure. This is a lever of the third order, FPW. In backward bending of the trunk, the lever and fulcrum and weight are the same; the power is applied at the front of the pelvis—lever of the third order, FPW. In both cases, the power-arm is fairly long, and although the weight-arm is greater than the power-arm, the leverage mechanism is efficient.

From the lying position, leverage action, at the hip joint, on either lower extremity, is a different problem (Fig. 12a). The lever is the entire limb—thigh, leg and foot. The fulcrum is at the hip joint, the power is applied to the shaft of the femur in front, the weight is applied to the center of gravity of the moving part (situated somewhat above the knee). It is a lever of the third order, FPW. If the knee is flexed concurrently with hip flexion, the center of gravity of the moving segment is shifted to a point much nearer the fulcrum, requiring less power to flex the limb and thus to move the lever. (As definite contraction of the abdominal muscles are required to fix the pelvis—to keep it from rotating downward—the movement, limb raising from lying position, is often used as an abdominal muscle developer. This is especially valuable in lying, double thighs-legs-feet raising).

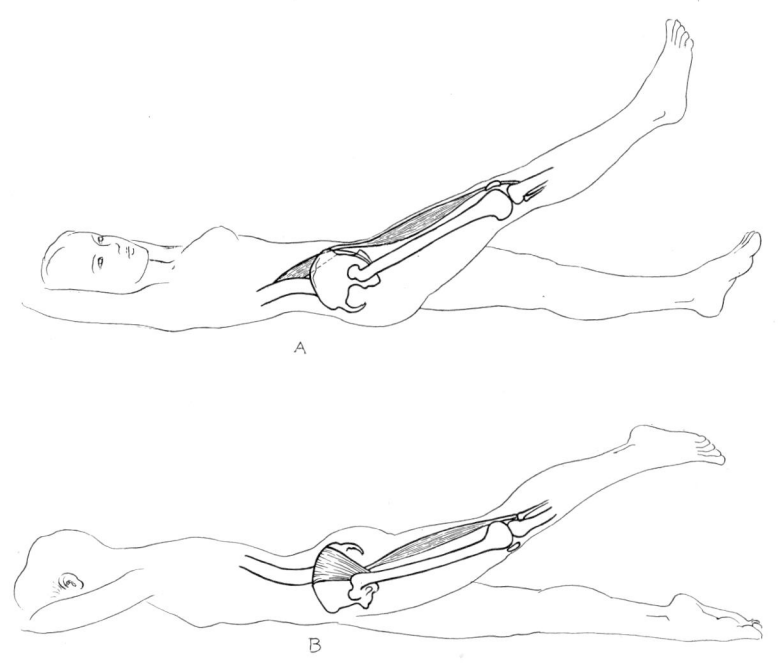

Fig. 12. Muscle mechanism in hip actions, about ⅕ life.
A. Right hip flexion from lying position.
B. Left hip extension from prone lying.

In prone lying, limb raising (Fig. 12b), the leverage is similar to the preceding. The lever is the entire lower extremity, the power is applied to the shaft of the femur in the rear—most of it by the Gluteal muscles and nearer the fulcrum than are the power applications in front—, the weight is the center of gravity of the limb. It is a lever of the third order—FPW. Here, too, if the knee is first flexed, the center of gravity of the moving part is shifted much nearer to the fulcrum and the power necessary to raise the limb is much less.

(This exercise is also used as a developer of low back muscles, as their service is required to prevent upward rotation of the pelvis).

From half-prone lying on a table, with feet touching the floor, single and double thigh raisings are used as scoliosis corrective and back muscle developers (Fig. 84, p. 226).

RECIPROCAL ACTIONS OF MUSCLES PASSING OVER MORE THAN ONE JOINT

For effectiveness in bodily activity, some muscles have a double (and occasionally multiple) action over joints. Thus, the calf muscles, acting on the knee joint and on the ankle joint, flex the knee and plantar flex the foot (Fig. 8, p. 26). This double action gives more power and purpose to walking, running and leaping. Also, the strong Quadriceps in the front of the thigh, acting on the hip joint and on the knee joint, first raises the thigh from the ground and then extends the leg in walking, running and leaping. The Hamstring muscles (Biceps femoris, Semitendonosis and Semimembranosis) at the rear of the thigh, first push the trunk forward by extension at the hip joint and then raise the leg (and the foot) from the ground, in acts of locomotion. Incidentally, this group of muscles is the mechanical factor in an important diagnostic test for low back strain. (Goldthwaite's sign). The Psoas magnus, passing over the low vertebral joints and the hip joint, flexes the trunk on the pelvis, acting from below upward, and flexes the thigh on the pelvis, acting from above downward (Fig. 12a, p. 41). This combined action assists in trunk raising from a lying position.

KINESIOLOGY OF THE HIP JOINT

To maintain posture balance in Fundamental Standing Position, tonic muscular contractions are larger factors at the hip joints than at either of the other major joints so far discussed. That is, we have seen that, at the ankle, there are but two motions involved, forward swaying (dorsi flexion) and backward swaying (plantar flexion), the mortise-like grip of the malleoli on the astragalus aiding greatly in ankle stability; and that, at the knee, the hazard to posture lies only in the prevention of forward "give" (flexion), as the strong crucial and rear capsular ligaments hold the knees from backward "give" (over extension); whereas nothing but muscular action prevents any of the possible motions of the hip joints—flexion, extension, adduction, abduction, inward rotation and outward rotation—except circumduction. None of these six motions have ligamentous or bony limitors until the body may have taken many degrees of motion in any direction. However, in good posture, the line of gravity of the body passes practically through the hip joints, viewed laterally, and thus minimizes forward bending of the trunk (hip flexion) and backward bending (hip extension); and viewed from the rear (or front) passes midway between the hip joints and so lessens the tendency to lateral trunk swayings (abductions and adductions at the hip) and to trunk twistings (inrotations and out rotations at the hip). Still, in assuming and maintaining Fundamental Standing Position, continued tonic muscular contraction is necessary. So we note that to prevent forward bending of the trunk, increased activity of the hip joint extensor muscles are involved; to prevent backward bending, the flexors of the hip are used; to prevent body swaying to the right (Fig. 13)—adduction at the right hip and abduction at the left hip—the abductors of the right and the adductors of the left hip joints come into play, and finally to prevent trunk twisting to the right—inrotation of the right hip and outrotation of the left hip—the outrotator muscles of the right hip joint and the inrotators of the left hip joint are the restraining factors.

RECIPROCAL MOVEMENTS AND LEVERAGE PROBLEMS INVOLVING BOTH HIPS AND BOTH ANKLES IN FUNDAMENTAL STANDING POSITION

1. In standing, body swaying to the right, there is a combination of (Fig. 13):
   a. Abduction at the right ankle—lever of the third order, FPW: The lever is the right leg and thigh, fulcrum at ankle joint, power by ankle abductor muscles, applied on the other side of the leg, weight far above;

44   BODY POISE

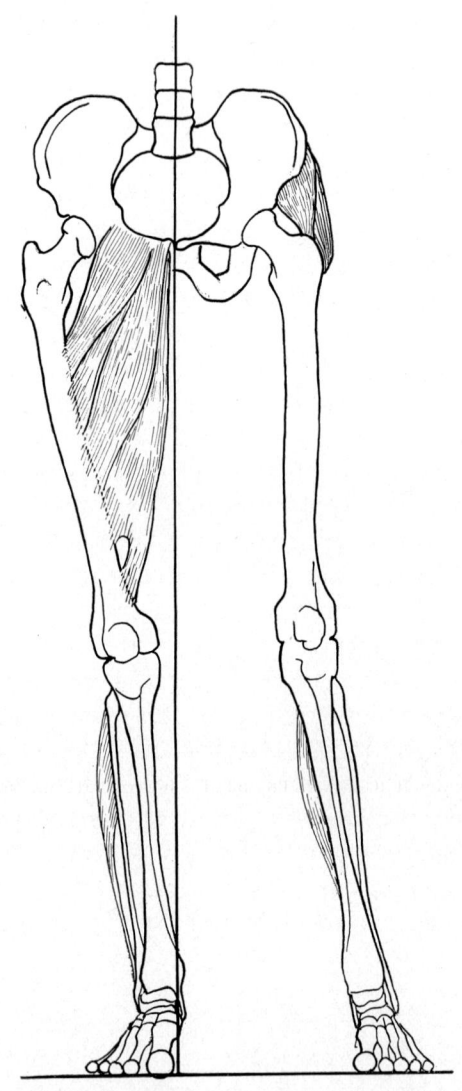

FIG. 13. Reciprocal joint and muscle mechanism, at hips and ankles, in standing, body swaying to the right, $\frac{1}{8}$ life.

    b. *Adduction at the left ankle*—Lever of the third order, FPW: the lever is the left leg and thigh, power by ankle adductors, applied on the inner side of the leg, weight far above;

c. Adduction at right hip—Lever of the third order, FPW: the lever is the pelvis, the fulcrum is the right hip joint, the power is the right thigh adductors, applied to the lower portion of the pelvis, the weight (or resistance) is toward the other side of the pelvis, at the opposite hip joint;

d. Abduction at the left hip—Lever of the first order, PFW: the lever is the pelvis, the fulcrum is the left hip joint, the power is the left thigh abductors applied to the outside and the flanges of the pelvis, the weight (or resistance) is at the opposite hip joint.

Similar reciprocal movements and leverage problems of these four joints, in Fundamental Standing Position, can be worked out for body swayings forward and backward and for twistings of the frame involving rotations at the hips and pronations and supinations of the feet. (It will be remembered that twistings are not possible at the ankles, but the rotation results, here suggested, are attained between the tarsal bones of the feet). The levers involved are all of the third order.

THE LUMBAR SPINE

Five superimposed vertebrae, the first, second, third, fourth and fifth, constitute the lumbar spine (Figs. 14a and b). Each vertebra has a body and a neural arch, which latter encloses a neural canal for the transmission of the spinal cord and its membranes and blood supply. The bodies are massive and puck-like in shape and each upper one is slightly less bulky than its neighbor below. The body is flattened above and below for the placement of the inter-vertebral cartilages. The neural arch is divided, for description, into two pedicles, two laminae, two transverse processes, four articular processes (two upper and two lower) and one spinous process. Together these portions form a ring for the encirclement of the spinal canal. The transverse processes and the spinous process afford leverage for the attachment of muscles and added protection of the spinal cord. The direction of the articular processes—slightly backward and mostly inward for the upper pair and slightly forward and mostly outward for the lower pair—regulates the movement of the spine in this region.

46   BODY POISE

Below each pair of pedicles is a groove for the transmission of spinal nerves. The lumbar spine, as a whole, has a normal moderate anteriorly directed curve. This curve is possible because the frontal vertical measurement of the bodies is slightly greater than the rear vertical measurement of the bodies.

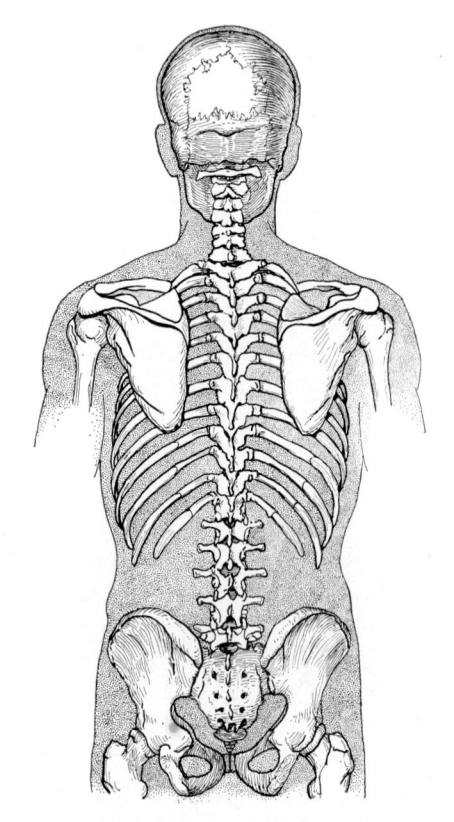

Fig. 14A. Bones of the head, spine, ribs, scapulae and pelvis, rear view, ⅑ life

Also, it is of interest to note that, besides man, this normal forward curving of the lumbar spine is found only in the higher apes, or only in those animals which approach erect carriage. The normal curves of the spine—the thoracic and the cervical as well as the lumbar—mitigate the harmful effects of jarring in undue activities of the individual. There are no statistics indicating the normal amount of forward curve in the lumbar spine or backward curve in the thoracic spine; i.e., what ratio of the ordinates to the chords of these curves is within physiological limits. We note however that, generally speaking, increase in forward carried belly increases the amount of forward curvature of the lumbar spine—hollow back—, and that faulty habits of

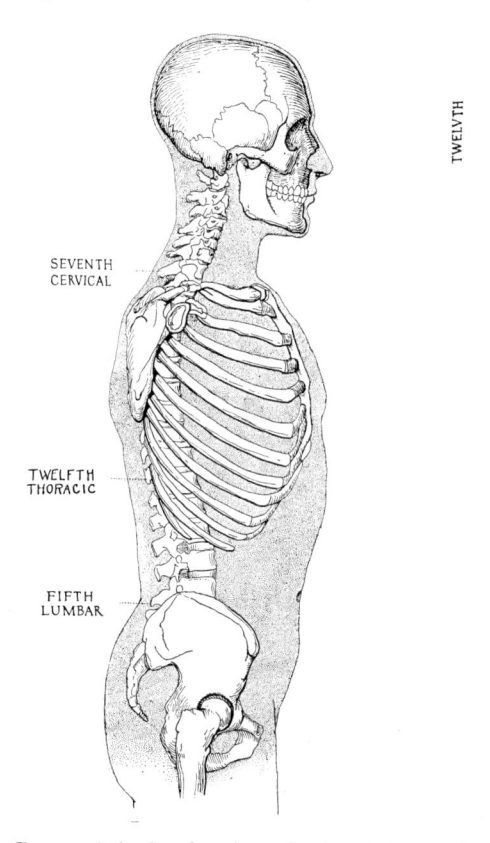

FIG. 14B. Bones of the head, spine, chest, pelvis, right side, ⅑ life

posture increase the amount of backward curvature of the thoracic spine—round back.

All of the bony parts of the vertebrae, except the pedicles, afford attachment to ligaments and muscles. The *ligaments* attached to the lumbar spine are, the intervertebral cartilages, interarticular cartilages, anterior common, posterior common, lateral common (in pairs), capsular (in pairs, around the articular processes), ligamentum subflava (in pairs), supraspinous and interspinous. The lumbar spine affords *origin* to Psoas magnus, Latissimus dorsi, Transversalis, Quadratus lumborum, Erector spinae and the deeper shorter muscles of the back and to the diaphragm, and *insertion* to Quadratus lumborum and the deeper shorter muscles of the back.

The mechanical function of the lumbar spine is primarily support to the superstructure and partial support and protection to the abdominal contents. It is also a part of the leverage mechanism of the trunk, in many human activities.

THE SACRO-LUMBAR JOINT

This is an amphiarthrodial or mixed joint. The body of lumbar V is superimposed upon the body of the sacrum, with the interposition of an elastic fibrocartilage, and there is articulation between the pairs of articular processes of lumbar V and the articular processes of the sacrum. The bodies of these vertebrae and the intervertebral discs bear the weight of the superimposed body segments and allow some motion; the articular processes bear a minimum of body weight but are all-important in giving direction and limitation to the movements of this joint. (These articular processes are subject to certain congenital anatomical variations, which may cause pathological symptoms).

The Ligaments of the Sacro-lumbar Joint

a) The intervertebral disc, a tough elastic substance placed between the bodies of the vertebrae, larger and thicker at this than at any other spinal joint; b) Anterior common ligament, connecting the fronts of the bodies of the vertebrae; c) Posterior common ligament, found within the spinal canal and connecting the rears of the bodies; d) Capsular ligaments, in pairs, completely surrounding the articular joints; e) The ligamenta subflava, in pairs, uniting the bases of the neural arches of the consecutive bones; f) Supraspinous ligaments, passing from tip of spinous process to tip of spinous process; g) Interspinous ligament, connecting the shafts of consecutive spinous processes; and h) Ilio-lumbar ligaments, in pairs, connecting the body of lumbar V with the crests of each ilium.

The Movements and Muscles of the Sacro-lumbar Joint

These are in all directions. However, owing to the limitations of compressibility of the intervertebral discs, but especially to the shapes and directions of the articular processes, movements are not great in any direction. (Note: This statement of comparative limitation of movement is true of all vertebral joints; but in the spine, considered as a whole, total movement is considerable). a) *Flexion* is *caused* by Rectus abdominis, Obliquus externus, Obliquus internus muscles, acting from the pelvis on the bony thorax. Flexion is *limited* by ligaments connecting the pelvis with the low spine, at the rear. b) *Extension* is *caused* by Latissimus dorsi, Erector spinae muscles, and the lower segments of some of the deep muscles of the back. Extension is *limited* by the strong anterior common ligament and the sacro-lumbar ligaments and by the approximation of the spinous process of lumbar V and the upper sacral spinous process and by the impinging of both articular processes of lumbar V vertebra on the articular process of the sacrum. c) *Lateral Motion* is *caused* by Obliquus externus, Obliquus internus, Quadratus lumborum, Erector spinae and the lower segments of some of the deep muscles of the back, all acting on the side toward which flexion is made. Lateral flexion is *limited* by the ligaments on the side opposite and by the approximation of the transverse process of lumbar V with the crest of the ilium. (Note: Habitual displacement, at this joint, involving continuous pressure of the transverse process of lumbar V on the crest of the ilium, is one of the recognized factors in low back pain). d) *Rotation* to a slight degree is also possible at this joint. It is *caused* by unilateral action of several of the muscles heretofore named and acting on the side toward which rotation is made—Latissimus dorsi, Obliquus externus, Obliquus internus—and by the Psoas magnus of the side opposite to the motion. Rotation is *limited* by practically all sacro-lumbar and ilio-lumbar ligaments and by the impinging of the articular process of lumbar V, of the side of motion, on the corresponding sacral articular process. (Note: The direction of the articular processes of all of the vertebrae, especially the manner in which articular surfaces contact each other, regulates the directions of motions and the limitations of motions, in the joints involved).

KINESIOLOGY OF THE SACRO-LUMBAR JOINT

In normal spines, the bony structure and the ligaments involved in the joint, are efficient in maintaining good posture. But the muscles controlling the joint do their part. However, in the discussion of abnormalities in this region, it will be found that the sacro-lumbar joint is an important factor in good posture and in functional disability. (Forward shifting, spondylolithesis, and lateral shifting of lumbar V on the sacrum are pathological conditions which may cause considerable derangement of the spine above).

JOINTS OF THE LUMBAR SPINE

Four joints are here included, as the sacro-lumbar has been described and the dorso-lumbar will be discussed separately. These are amphiarthrodial or mixed joints. The body of each vertebra involved rests on its neighbor below, with the interposition of a fibro-cartilage; and between each two pairs of articular processes of one vertebra, there is a joint with the articular processes of the vertebra below. The superimposed weight factor is an ever, though slightly decreasing, factor as higher and higher spinal joints are encountered. Motions in these joints are those of flexion, extension and lateral flexion, but not of rotation. The laterally faced articular facets prevent rotation in the lumbar spine. All motions are centered about the middle of the intervertebral discs, which might be likened to compressible rubber pads. With any motion, the side of the disc toward which motion is made, is compressed and the opposite side somewhat expanded. But, as has been stated, the kind and amount of motion is directed and limited by the facing directions and positions of the connecting articular processes. It is this combination of universally directed movement between bodies of the vertebrae and the vertebral fibro-cartilages, on the one hand and between the pairs of regulating articular processes, on the other, that gives to spinal joints the name of mixed joints.

The ligaments of the lumbar joints

These are intervertebral fibro-cartilages (each about one-fourth of the thickness of the neighboring vertebral body) interposed between consecutive vertebral bodies; the interarticular fibro-cartilage, found between the faces of the joining articular processes, which are smooth, slightly elastic and facilitating the sliding movement of these faces; the anterior common ligament, connecting the fronts of the bodies of the lumbar (and indeed all of the spinal) vertebrae; the posterior common ligament and the lateral common ligaments (in pair) connecting the rear and the sides respectively of the bodies of the vertebrae; the capsular ligaments (in pairs) surrounding the entire articular faces of the articular processes; the ligamenta subflavae (in pairs) connecting the laminae, and the interspinous and the supraspinous ligaments, connecting the shafts and the tips of the spinous processes.

The Movements and Muscles of the Lumbar Joints

Like the sacro-iliac joint, because of compressibility of the intervertebral discs and the direction-governing articular processes, there is but little motion between any two of these joints; but in all four there is a summation of motion which is not inconsiderable.

a) *Flexion* is *caused* by Rectus abdominis, Obliquus externus, Obliquus internus and by the Psoas magnus muscles, acting from the pelvis on the thorax. Flexion is *limited* by the ligaments in the rear of the spine and by the compressibility of the intervertebral discs. (Note: Although obesity is not a normal condition, the presence of abdominal fat or any pathological growth in the abdomen causes limitation of flexion).

b) *Extension* is *caused* by Latissimus dorsi, Erector spinae, and the lower segments of some of the deep muscles of the back. Extension is *limited* by the strong anterior common ligament and by the approximation of the spinous processes each upon its neighbor.

c) *Lateral Motion* is *caused*, in either direction, by the Obliquus externus, Obliquus internus, Quadratus lumborum, Erector spinae muscles—all acting on the side toward which motion is made. Lateral flexion is *limited* by the ligaments and, to some extent, by the tightening of the muscles, of the side opposite to that of the motion, and by the approximation of the transverse processes of the same side. At the limit of lateral motion, tightening of the capsular ligament, of the same side, is a factor.

KINESIOLOGY OF THE LUMBAR JOINTS

From the viewpoint of Fundamental Standing Position, this is the next body segment in the "up-building" of Posture. At the sacro-lumbar joint, in normal spines, the bony structure and the ligaments involved in this region go a long way toward maintaining good posture; but with the increase in possible "breaks"— four in this segment—there is ever greater hazard to good body poise. The aid of muscles is here called upon to a greater extent than in the single joint below. The normal movements away from the stable position are caused by gravity and by the muscles on the side of the movement, and are prevented by the tonicity of the opposite muscles. (This has been sufficiently discussed in the previous paragraph. The "slouchings" possible in this region, what causes them and what prevents them, will be discussed in the chapter on abnormal posture).

LEVERAGE IN THE LUMBAR SPINE AND JOINTS

In standing position, leverage principles are involved in forward bending (flexion), backward bending (extension) and side bending (lateral flexion) of all of the superstructure above the pelvis (sacrum). Study of each of these motions could be made as they involve seriatim, each of the joints included—that is, lumbar V and all above it on the sacrum, lumbar IV on lumbar V, etc. But the principles differ only in the fact that the lever-arm and therefore the Moment of Weight is a trifle less for each upward following joint. We will therefore discuss the problems of variance as at any of the lumbar joints. In all motions, the lever-arm is the spine from lumbar IV upward. The fulcrum is the intervertebral disc. The weight is applied at the center of gravity of the superimposed segment; the power varies with the motion:

a) In trunk bending forward, the power is applied at the front of the lumbar and thoracic vertebrae and of the lower chest wall; b) In trunk bending backward, the power is applied at the rear of the lumbar and thoracic vertebrae and of the lower chest wall, and even to the upper end of the humerus (Latissimus dorsi action); c) In trunk bending sideways, the power is applied at the sides of the lumbar and thoracic vertebrae and of the lower chest wall opposite to the direction of motion.

Each of these leverages is of the third order, FPW.

THE THORACIC SPINE

Twelve vertebrae—Thoracic I to Thoracic XII inclusive—comprise the thoracic spine. As in the lumbar spine, each vertebra is superimposed upon its neighbor below—Thoracic I, upon which rests the seventh cervical vertebra, at the top, and Thoracic XII at the bottom, resting upon Lumbar I. From lowest to uppermost, each vertebra is slightly smaller than that below it. The twelve are so placed as to form a gentle curve bulging backward, the sixth thoracic being placed farthest backward. As in the lumbar region, each thoracic vertebra has a body and a neural arch and the latter completes a neural canal. The bodies are less massive than those of the lumbar region but of larger bulk than those found in the cervical region. Above and below each body is a flattened surface for contact with the intervertebral fibro-cartilages. At the side of each body, near its pedicle, is an articular facet for the articulation of the corresponding rib. Here, too, each neural arch is divided into two pedicles, two laminae, two transverse processes, four articular processes and one spinous process. The ring which the neural arch completes, affords space for the transmission of the thoracic portion of the spinal cord. The transverse processes give attachment to ligaments and muscles and protection to the spine. Near the tip of each transverse process is an added facet for further articulation with a rib. The articular facets, in two pairs, face respectively upward backward and downward forward, allowing flexion, extension, lateral motion and rotation. The spinous processes are less massive than those of the lumbar region but are longer and extend backward downward. On either side of the spinous processes and between them and the pairs of transverse processes are deep grooves for the attachment of extensor and rotation muscles of the spine. The transverse processes and the spinous processes afford considerable leverage-arms for the attachment of muscles. As in the lumbar region, below each pair of pedicles are grooves for the transmission of spinal nerves. The thoracic spine as a whole has a normal posterially directed curve, beginning at the thoracico-lumbar joint, greatest at the fifth or sixth thoracic vertebra and ending at the cervico-thoracic joint. The front perpendicular lengths of the vertebrae are less than the corresponding rear measurements of each bone. This discrepancy is very little in any two bones but amounts to a considerable arc in the total of the twelve bones. All bony parts of the vertebra, except the pedicles, afford attachment to ligaments and to muscles.

The *ligaments* are the intervertebral discs, interarticular cartilages, anterior common, posterior common, lateral common (in pairs), capsular (in pairs around the articular processes), ligamenta subflavae (in pairs), supraspinous and interspinous. In addition, certain ligaments are found at the junction of the rear of each rib with its neighboring vertebra. They are capsular, interarticular and stellate, where the extreme rear of the rib joins the body of the vertebra, and capsular, superior costo-transverse, middle costo-transverse and posterior costo-transverse, found where the so-called neck of the rib articulates with the tip of the transverse process.

The *muscles* attached to the thoracic spine are *origin* of Trapezius, Rhomboids and Latissimus dorsi, acting on the upper extremity, of Erector spinae, Spinalis dorsi, Semispinalis dorsi, Multifidus spinae, Rotatores spinae, extending to other vertebrae above, of Levatores costarum, Serratus posticus superior, Serratus posticus inferior, extending to the ribs, Splenius capitus, Splenius colli, Tracelo-mastoid, Complexus, Semispinalis dorsi, Semispinalis colli, extending to neck or head or both. The thoracic spine gives *insertion* to Latissimus dorsi, Spinalis dorsi, Semispinalis dorsi, Multifidus spinae, Rotatores spinae muscles, acting from below, and to Splenius coli, Semispinalis dorsi, acting from above.

The mechanical function of the thoracic spine is support to the upper torso, to the head and neck, special support to the bony thorax and through this segment to the arms; also protection to the thoracic and the spinal canal contents.

THE LUMBO-THORACIC JOINT

This joint is considered separately because here the spinal column, as a whole, is subject to more stress and strain than is found either above or below it. This increased stress is due, in part, to the fact that the joint marks, in its bony mechanism, a transition between the characteristics of the lumbar spine and those of the thoracic spine. But the unusual strain is due more particularly to the fact that, at this point, the segment of the massive lumbar column joins the lighter but reinforced thoracic column. This thoracic segment, composed of the thoracic spine and the bony thorax, is also massive and compact, but of a different character from the single stemmed lumbar segment. This element of weakness might be illustrated by an occasional experience of a tree-nursery man. A sapling is set out and, because it is immature, its long thin trunk is reinforced by parallel stakes to which it is tied, up to the spreading branches. The mass of spreading branches more or less protects itself. But in a hurricane, the trunk is snapped at the upper limits of the reinforcing stakes. So, as this comparative joint leads not infrequently to static disabilities, separate consideration of the lumbo-thoracic joint is pertinent.

The *Lumbo-thoracic Joint* is an amphiarthrodial or mixed joint. The body of Thoracic XII rests upon that of Lumbar I, with the interposition of fibrocartilage; and there are a pair of articular joints between the articular processes of the two bones. The articular surfaces of the latter face each other in planes that are midway between the side to side arrangement in the lumbar vertebrae, which prevented rotation there, and the backward upward and forward downward direction of the typical thoracic vertebrae. This arrangement allows a moderate degree of rotation but not as much as is possible in the thoracic spine. The motions at this joint are flexion, extension, lateral flexion and limited rotation. The *ligaments* and *muscles* are the same as in the lumbar spine. Of the *movements* at this joint, nothing is to be added concerning flexion, extension and lateral motion to what was described in the movements of the lumbar spine. Also, the discussion of the anatomy of rotation will be reserved to that pertaining to the thoracic spine.

KINESIOLOGY OF THE LUMBO-THORACIC JOINT

This deserves attention. In the first place, the addition of rotation to the normal movements of the spinal column, presents an increased onus to the muscular system; and this segmental weakness (mentioned above) due to the "break" between the massive lumbar spine and the massive thoracic segment—thoracic spine plus bony chest—is a distinct "hazard" to the maintenance of good posture. As has been said, good posture, in the standing position, depends upon the habitual maintenance of more or less equally distributed body weight around the line of gravity of the body. Therefore presupposing the fixation (by means which have been pointed out) of the body structure from the soles of the feet to the thoracicolumbar joint, postural variation of the body segments, above that joint, involves combating the force of gravity upon such upper segments, in the angular motions of flexion, extension and lateral flexions, and overcoming the muscular or other forces which would (unobstructed) cause rotation. To meet these needs, the shape and position of the parts of the component bones and the ligaments do their share, but muscles must exert greater influence in this region than either above or below it. Thus, flexion is prevented by the tonicity of the muscles of the back (Erector spinae, Multifidus spinae, etc.); extension is maintained by the abdominal and other muscles at the front; lateral motion is prevented by the tonicity of the muscles of the opposite side, and rotation to either side is prevented by bringing into play the rotation muscles of the other side.

THE THORACIC JOINTS

Eleven amphiarthrodial or mixed joints are here discussed (the cervico-thoracic joint to be considered later). They do not differ from other spinal joints except in the movement-control due to the interfacing of the articular facets and to the fact that the intimate relation of the ribs to all of these joints limits all motions in this region.

The *ligaments* bind bones to bones as in the spinal joints below, and there are similar interarticular fibro-cartilages between bodies and articular cartilages lining the articular facets.

The Movements and Muscles of the Thoracic Joints

Like the lumbar joints, the centers of motion are in the compressible interarticular fibro-cartilage and, if unrestricted by other factors, this rocking motion would be as free here as elsewhere in the spine. But there are decidedly limiting factors, mostly bony, in the thoracic region.

a) *Flexion* is *caused* by Rectus abdominis, Obliquus externus and Obliquus internus muscles, acting from the pelvis on the lower chest and through the chest on the thoracic spine. Flexion is very *limited*, due to the approximation of the ribs upon each other and to the direction of the faces of the articular processes.

b) *Extension* is *caused* by the Erector spinae and Multifidus spinae muscles, acting from the pelvis on the thoracic spine, by Semispinalis dorsi and by the Rotatores spinae muscles, acting from lower on higher thoracic vertebrae and by Semispinalis dorsi, acting from the upper thoracic spine on the neck. Extension is *limited* by approximation of the spinous processes on each other and by resistance of the chest wall. (It will be seen, however, when discussing the movements of the chest, that extension of the thoracic spine is a real aid to elevation of the ribs and depression of the diaphragm in increasing chest capacity in the physiological function of respiratory inspiration.)

c) *Lateral Flexion* is *caused* by muscles of the side toward which motion is made—i.e., by Erector spinae, acting from the pelvis on the thoracic spine, by Quadratus lumborum, Obliquus externus and Obliquus internus, acting from the pelvis on the lower ribs, by Semispinalis dorsi, Multifidus spinae and by Rotatores spinae, acting from lower on higher thoracic vertebrae. Lateral flexion is somewhat freer than flexion or extension, but this movement is also *limited* by resistance of the chest wall.

d) *Rotation* is fairly free in the thoracic spine, especially in the upper part. Some of the muscles twist this segment to the same side, others to the opposite side. Those which rotate to the same side extend from a centrally fixed lower to a laterally higher insertion. They are Latissimus dorsi, from pelvis, across many joints, to the humerus, and the Obliquus internus, acting from the pelvis on the lower ribs. Those which rotate the thoracic spine to the opposite side act from a lower lateral to a higher central insertion. They are, Multifidus spinae, Rotatores spinae, extending from thoracic vertebrae and Obliquus externus from pelvis to ribs.

KINESIOLOGY OF THE THORACIC SEGMENT

As an entity in the human skeleton, this segment, composed of twelve thoracic vertebrae, is comparatively immovable. Therefore possible displacements of the segment (plus head, neck and upper extremities) will here be considered as if taking place at the thoracico-lumbar joint. In the first place, we note that the line of gravity of the body, in standing position, passes through this segment well to its rear. All lateral motions are instigated by muscles on the side toward which motion is made and carried on by gravity, and are prevented or regulated by the tonicity of the muscles on the side contra-lateral to that of the movement. (It is to be noted that when stated exercises, derived from the Fundamental Standing Position, are given for specific muscle development, it is the opposite or steadying muscles which are improved. That is, these are the muscles which act on the moving segment.)

a) *Forward bending* of the thoracic spine is initiated by the muscles in front of the body (abdominal and other) and carried on by gravity; but is steadied by the eccentric action of the muscles of the back (Erector spinae and others).

b) *Backward bending* is initiated by the muscles of the back (Erector spinae and others) and carried on by gravity and is regulated by the eccentric action of the abdominal muscles.

c) *Side bending* is initiated by the muscles of the side of the movement (abdominal muscles, Quadratus lumborum, Erector spinae and others) and is carried on by gravity, but is controlled by the eccentric action of the muscles of the contra-lateral side.

In (d) *Rotation*, the mechanical principles are different. Gravity plays little or no part—the movement is kept largely within and about the line of gravity. With this motion, muscles act concentrically and directly on the moving segment or subsegments of the region. However, because some of the rotating muscles act in front of the line of gravity axis, while others have their acting attachment to the rear of this axis, some of the muscles, on the side of the motion, turn the moving segment to the same side while others turn to the opposite side. And the opposite groups simply regulate or limit such motion. Thus, in standing trunk-twisting to the right, there is direct action of Obliquus internus (pelvis to ribs) and of Latissimus dorsi muscles (pelvis to humerus) on the right side; and there is action of Obliquus externus, of Multifidus spinae and of Rotatores spinae of the left side. (Note: Obliquus externus is situated mostly in front of the turning axis, but turns the segment to the opposite side because, from its pelvic or fixed attachment, its course is upward outward and backward and applied to the ribs back of the turning point. Thus, in thoracic rotation to the right, Obliquus externus of the left side is used.)

LEVERAGE OF THE LUMBO-THORACIC SPINES AND JOINTS

Leverage, for the angular movements, does not differ in principle from that described for the lumbar region. All of these levers are of the third order, FPW. Rotation, however, presents a new problem. Gravity plays but little part. In trunk-twisting to the right, for instance, the fulcrum is the intervertebral disc; the lever is the right side of thoracic vertebra XII, the chest wall of the right side and even the scapula and humerus (because Latissimus dorsi is one of the moving muscles); the power is applied extensively to the transverse process of vertebra XII, to the chest wall and to the upper end of the humerus; the weight (resistance) lies beyond the point of application of the power, considering the short muscles, and making this a lever of the third order, FPW, and between the fulcrum and power, where the action of the Latissimus dorsi is considered. This leverage is therefore one of the second order, FWP.

THE CHEST—BONY THORAX (FIG. 15)

The chest is a large cage, holding the heart, the principal blood carrying stems, the lungs and the oesophagus, and having definite movements necessary to the mechanism of breathing. It consists of twelve pairs of ribs, special fibro-cartilages and the sternum or

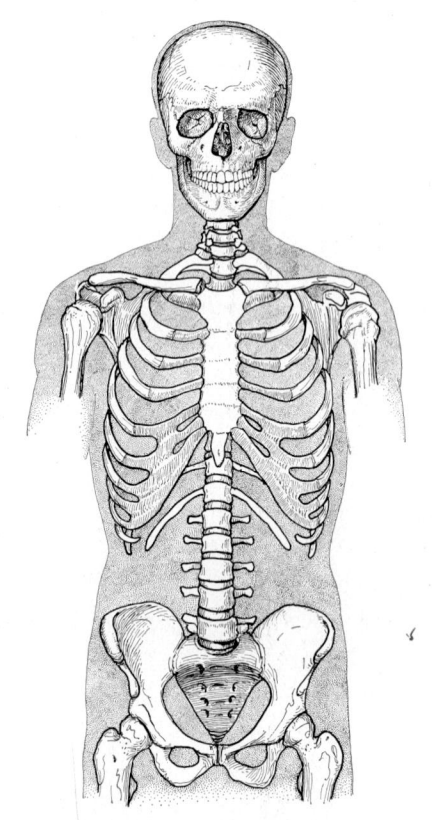

FIG. 15. Bones of the head, spine, chest, pelvis, front view, ⅓ life

breast bone. The *ribs* of each side of the chest are divided into seven true ribs and five false ribs. The false ribs are sub-divided into three costo-chondral and two floating. All ribs articulate with the thoracic spine, at the rear. The seven true ribs join in front to the sternum, by the interposition, in front, of its costal cartilage; the cartilages of the next three (costo-chondral) unite each with the costal cartilage of the rib next above, and the last two (floating ribs) have each a short costal cartilage but no chondral or bony union in front. Thus, all ribs are made up of a greater portion of curved flattened bone at the sides and rear of the chest, and of fibro-cartilage in front. We will see that this elastic fibro-cartilage plays an important part in the chest movements of respiration.

EACH RIB is divided for description into a head, at the rear, a neck and a shaft. All ribs leave the spine in a direction outward and a little backward, then curve in an ovoid shape forward and downward, around the chest contents, and finally, (except the floating ribs) forward centerward. From above downward, the curved length of each rib increases to the seventh; thereafter the length decreases until the twelfth is nearly as short as the first.

The STERNUM or breast bone completes the front boundary of the bony chest. It is a thin, flat bone, about four times as long as it is wide. Because of a fancied resemblance to a gladiator's sword, it is divided into a manubrium or handle, a gladiolus or blade and a xiphoid appendix or point.

The resulting *shape* of the *bony thorax* as a whole is something between a cone and a dome, with the base at the bottom and the apex at the top. The bottom is closed by the Diaphragm (to be described later), except for the transmission of the oesophagus, important circulatory and nerve trunks and the upper portion of Psoas magnus muscle. The lateral diameters of the chest are greater than the antero-posterior.

Ligaments of the Bony Thorax

The bony thorax affords attachment to ligaments and muscles, both outside and inside. The ligaments attached to the ribs, at the rear, are the capsular, interarticular and stellate or anterior costo-vertebral, joining ribs to bodies of vertebra; the capsular, superior costo-transverse, middle costo-transverse and posterior costo-transverse, joining the ribs to the transverse processes; and in front, the anterior chondro-sternal, posterior chondro-sternal, superior chondro-sternal, inferior chondro-sternal and an occasional interarticular ligament, joining the costal cartilages with the sternum or (as in some of the lower ribs) the cartilages each with that above, and finally there is a chondro-xiphoid ligament which unites the xiphoid appendage with the sternum.

Muscles and Movements of the Bony Thorax

The muscles attached to the ribs, sternum and the Xiphoid are found mostly on its outer wall and on the lower rim of the chest; but some are located wholly within the bony thorax. They are *origin* of Obliquus externus, acting on abdomen and thus on lower body segments; of Sterno-mastoid, Sterno-hyoid, Scalenus anticus, Scalenus medius, Scalenus posticus, Cervicales ascendens, acting on head or neck; of Diaphragm, External intercostals, and Internal intercostals, acting from ribs to ribs; of Pectoralis major, Subclavius and Serratus magnus, acting from chest wall to scapula or clavicle, and of Latissimus dorsi and Pectoralis major, acting from the chest to the humerus. The chest wall also affords *origin*, on its inner or visceral surface, to Triangularis sterni, acting from sternum to ribs. The bony thorax gives *insertion* to Rectus abdominis, Obliquus internus, Quadratus lumborum, Erector spinae, acting from the pelvis or lumbar spine; to Levatores costorum and Serratus posticus inferior, acting from the thoracic spine; to Serratus posticus superior, acting from the cervical spine or from the head, and of External intercostals, Internal intercostals, Infracostales and Triangularis sterni, acting from rib to rib.

JOINTS OF THE RIBS

The principal function of the chest is to protect the lungs, the heart, the large blood vessels and the oesophagus and to supply the movements necessary for breathing. One must clearly understand that actual intake of air into the lungs is the result of atmospheric pressure. The movements of the chest wall simply allow increased capacity for inflowing air through the nose, mouth and trachea by chest enlargement, on the one hand, and expulsion of air by lessening the chest capacity, on the other. In this section, we discuss the mechanism of this change of capacity of the chest. And we find that this increase, accompanying the function of inspiration involves increase in all diameters of the chest

—vertical, lateral and antero-posterior; and that the function of expiration involves decrease in all diameters. During inspiration, there is elevation and eversion of the ribs and depression of the Diaphragm, and during expiration, there is depression and inversion of the ribs and elevation or relaxation of the Diaphragm. The movement of the ribs, in inspiration, is a muscularly active (though usually unconscious) one of the entire chest wall. It includes a passive twisting of the rib cartilages in front. During expiration, depression and inversion of the ribs is a recoil or untwisting of these twisted cartilages, with inactivity of the inspiratory muscles, although certain small muscles are used to aid forced expiration. Also, it is to be noted that depression of the Diaphragm is an important part of the act of inspiration. Depression of the Diaphragm is an active movement. Its elevated central tendon is drawn down by an action pull of its encircling muscle fibres, working concentrically from the lower borders of the ribs and the low thoracic spine. This downward pull of the central tendon, this comparative flattening of the dome-like shape of the Diaphragm, involves pressure on the abdominal contents, with corresponding bulging of the muscular, fascial and dermal walls of the abdomen. Elevation of the Diaphragm (part of the of the expiratory phase of the mechanism of breathing) is a muscularly inactive function of the Diaphragm. Its muscle fibres simply "give"; but the recoil of the walls of the abdomen exert a mechanical pressure upon the abdominal contents, which in turn force the central tendon of the Diaphragm upward and so help to expel air from the lungs (fig. 16, p. 68).

JOINTS OF THE THORAX

The motions of the bony thorax are those of the physiological function of respiration, the mechanism of which will be discussed later. Here we study the intrinsic mechanism of the joints involved. They are divided into three groups: 1. costo-central; 2. costo-tranverse; 3. costo-sternal.

1. THE COSTO-CENTRAL JOINTS

Twenty-four in all, in pairs on each side of the spine. They are diarthrodial joints. The rear end or head of each rib has two articular facets, which join similar facets on the sides of the bodies of the corresponding vertebra (although, in some parts of the thoracic spine, the rib articulates in part with the intervertebral disc). The bones are held together by a capsular, an interarticular, and a stellate or costo-central ligament. The movement is a simple twisting of the rib head on the vertebral body. This motion is limited by the tension of the local ligaments and by the amount of twisting possible in the costal cartilages in front.

2. THE COSTO-TRANSVERSE JOINTS.

These are twenty-four in number, twelve on each side, and are lateral ginglymous or lateral hinged joints. The small tubercle, at the neck of each rib, has a facet which joins with a similar facet at the tip of the transverse process of the vertebra. The bones involved are held together by capsular, superior costo-transverse, middle costo-transverse and posterior costo-transverse ligaments. The movement is a lateral gliding of the movable head of a rib against a fixed transverse process. This motion is limited by the ligaments involved. This joint group gives added stability to the ribs, in the real movement, which takes place at the costo-central joints.

The *movements of these two joints*, costo-central and costo-transverse, should be considered together. It is really one movement. To clarify this, we may imagine the rib separated from its sternal attachment in front and then straightened in its entire length. The motion now would be a simple pivoting of the head of the rib against the side of the body of the spine (in some cases against the intervertebral fibro-cartilage); and the movement at the costo-transverse joint would be a ligament-guided lateral gliding. Or, the rib back in its normal shape and position and with the front reattached to the sternum, the single movement, of these two joints, is one of rib twisting about a linear axis extending from the costo-transverse joint to the center of the costo-central joint.

3. The COSTO-STERNAL JOINTS are ginglymoid or hinged joints. They are not uniform in the front attachment of all ribs. Although all ribs have fibro-cartilages at their anterior terminals, these fibro-cartilages are arranged as follows: the first seven extend from their respective ribs directly to the costo-sternal joints, with a gradual increase in length from the first to the seventh;

the next three extend each from the front terminal of the rib to the cartilage next above, allowing these three to have a pseudo-articulation, with the sternum, through the seventh costo-sternum articulation, and the last two, called floating ribs, have no frontal attachment. Thus it will be seen that there are but seven costo-sternal articulations on each side of the sternum or fourteen in all. The seven true ribs, at their cartilage ends, are held to the sternum by anterior chondro-sternal, posterior chondro-sternal, superior chondro-sternal and inferior chondro-sternal ligaments. The movement is a gliding and pivoting one. This motion is *limited* by the ligaments involved. If the seven ribs with their cartilage attachment could be detached from their rear joints and straightened, the movement at these joints would be also pivotal. (Note: Some text books on Anatomy add costo-chondral, chondro-chondral and intersternal joints, with corresponding ligaments. As the movements between ribs and cartilages, between cartilages and cartilages and between the parts of the sternum, are insignificant in the general study of body posture, they are here omitted.)

THE RESPIRATORY MOVEMENTS OF THE BONY THORAX AS A WHOLE

These are not simple. The inter-relation of spine, ribs, cartilage and sternum gives definite limitation to movement, but affords a remarkably efficient activity for a specific purpose—breathing. We will therefore discuss these two motions, which are named *elevation with eversion* and *depression with inversion* of the ribs. To describe this, we will consider a sectional ring composed, for example, of the fifth thoracic spine, the fifth rib on the right side, the fifth rib on the left side, two cartilages at the front of these ribs and a corresponding section of the sternum. This rather heart-shaped ring would have a general direction of forward and downward from its more fixed rear. The movement of elevation of this ring might be illustrated by considering the human trunk as the vertebral body, by placing a shingle (sternal segment) in one's hands and by elevating the arms (the ribs) away from the front of the human torso. Thus the movements at the illustrative human shoulder joints would correspond to those of the rear rib joints and, finally, likening the flexibility of the human wrists to the costo-sternal flexibility, we could portray the movements of the costo-carilages and the sternal articulations. To complete the simile, the elbows should extend laterally backward from the plane of the body (which, of course, would separate the hands and make the "sternal" shingle unduly long or we would be obliged to illustrate with the very long arms of apes). If now, we lift the shingle and proportionally lift the elbows, we would also widen the lateral and antero-posterior diameters of the open space inside the encircling part of our illustrative segment.

Eversion and inversion of the flat (though curved) ribs is illustrated by the opening and closing of the slats of a Venetian blind.

To recapitulate, the act of respiratory *inspiration* is accompanied by increase in all chest diameters and involves:

1. Extension of the thoracic spine;
2. Elevation and eversion of ribs, with twisting of the costo-sternal cartilages;
3. Depression and spreading of the floating ribs;
4. Depression of the Diaphragm.

1. Extension of the thoracic spine increases the vertical diameter of the chest cavity, by mechanically aiding the elevation of the upper six pair of ribs. (The muscles and limitations involved have been studied. See p. 62.)

2. Elevation and eversion of the ribs is *caused* by Levatores costarum, Serratus posticus superior, acting from the cervical and thoracic spine, by the External and Internal intercostal muscles, found between the margins of the shafts of each rib and acting from rib above to rib below. This action is *limited* by the costo-central and costo-transverse ligaments and by the torsion of the costal cartilages.

(Note: What has been called "air-hunger" is a real but temporary phenomena of breathing in active exercise. Physiologically, it is a need, on the part of the blood, for more oxygen, than is momentarily available, to meet the oxygen demand of increasingly active muscles all over the body. It is met by increased action of the respiratory muscles to enlarge the chest capacity and thus to allow more entry of air (oxygen) into the lungs. To those of sedentary habits, the air needs of the lungs are comparatively little. The normal balance between ultimate tissue needs for oxygen and the chest wall activities have been attained. With physically active persons, the habitual air needs are greater and the chest wall activities are greater, but with them the balance has been attained. But with sudden increase of activity, whether by those of sedentary or athletic habit, there is an almost immediate imbalance between the newly called-for tissue needs for more oxygen and the delayed increased function of the chest wall and Diaphragm. These early efforts of the chest wall mechanism are great, sometimes painful and with panting breath. But the balance soon is attained—more slowly in the sedentary, prompter in the athletically trained. The so-called "getting the second wind," essentially a bio-chemical mechanism, is expedited by adjustment of the chest-wall mechanism, by enlargement to meet its needs.

The usual, habitual chest excursion provides so-called "tidal air". Certain muscles are classified as aiding "forced" inspiration. They are the Scaleni, acting from the cervical and the thoracic spine on the upper ribs, and the Pectoralis major, Pectoralis minor and Serratus magnus, acting from otherwise fixed clavicles and scapulae.

3. Depression of the lower ribs is *caused* by Latissimus dorsi and Serratus posticus inferior muscles, extending from the lumbar and low thoracic spine to near the rear of the lower ribs. It is *limited* by the ligaments of the chest joints.

4. Depression of the Diaphragm. The Diaphragm is an important muscle of respiration. From an origin on the transverse processes of Lumbar vertebra II, the lower ribs and the xiphoid appendix of the Sternum—a complete though irregularly shaped circle—its muscular fibres extend upward and concentrically inward to a central tendon. It is thus dome-shaped, highest when most inactive, less rounded and lower with increasing muscular action. It forms the wall between the thoracic and the abdominal cavities, and is penetrated by the oesophagus and by important vessels and nerves. Upon it rest the lower borders of the lungs and the pericardium. In its concave space below are found the large liver, the stomach and the spleen.

The simultaneous action of the concentrically directed muscular fibres of the muscle causes descent of its central tendon. As this descent gives increase to the vertical diameter of the chest cavity, the Diaphragm becomes an important muscle factor in respira-

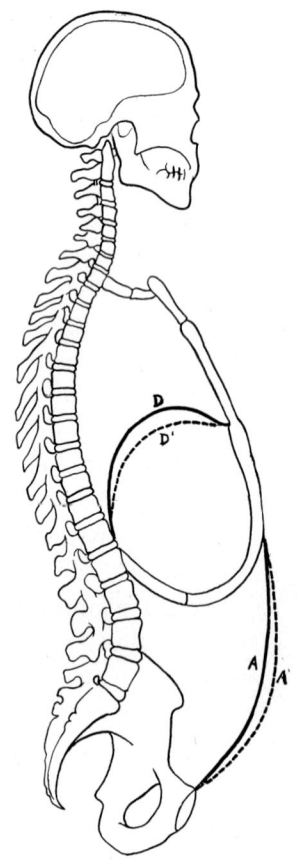

Fig. 16. Reciprocal mechanism of diaphragm and abdominal muscles in breathing, ⅛ life.
 D and A show approximate position in expiration.
 D' and A' show approximate position in inspiration.

tory inspiration. But it does more. It helps to prepare for the passive action of respiratory expiration, just as twisting of costal cartilages was seen to prepare for passive expiration .(Fig. 16) The descent of the Diaphragm compresses the abdominal contents. These contents must seek space. They bulge against the abdominal wall. (They can not affect the lumbar spine). We will see (below) that the recoil of these walls, when Diaphragm action ceases, gives reverse movement, in expiration. We can see the importance of moldable abdominal contents and of efficient abdominal muscles in insuring effectiveness of this mechanism of "abdominal breathing" (Abdominal tumors, including pregnancy and abdominal wall fat lessen this efficiency.) Thus we note that the mechanism of respiratory inspiration is active in every detail.

The act of respiratory *expiration* involves decrease in all chest cavity diameters and is for the most part a passive function. Decrease in all diameters is accomplished by release of the inspiratory muscles and the passive recoil of costal cartilages and of the abdominal contents—abdominal wall mechanism. The specific job, in expiration, is to force the deoxygenated air from the lungs through the upper air passages and so outside the body. This is a real physical task and the recoil mechanisms are practically adequate. But certain muscles aid in expiration. These are Triangularis sterni and Infracostales, acting specifically above and below on the front inside of the chest, and the abdominal muscles, acting by regaining tone after the stretch of the inspiratory act.

KINESIOLOGY AND LEVERAGE OF THE RIBS

In Fundamental Standing Position, the movements of the chest have but little effect upon the posture. The forward carriage of the sternum, increased in its antero-posterior diameter, in inspiration, shifts the center of gravity of the body slightly forward. Theoretically, this weight-shift would require slight readjustment of the entire human frame, from the ankles upward, to maintain balance. Practically, it is doubtful whether any such posture adjustment takes place.

Leverage problems in this region will be discussed as of:
1. Elevation and eversion of a single rib;
2. Movement of the bony thorax, as a whole, in inspiration; and
3. The complicated mechanism, due to the involvement of the chest with the upper spine, in rotation.

1. In *Elevation* and *Eversion* of the *fifth rib*, for instance, the rib is the lever; the fulcrum is at the axis of motion which passes through the head, the short neck and the tubercle of the rib; the power, Intercostal muscles from rib above, is applied along the curved course of the rib, and the weight (resistance) is the twisting of the costal cartilage and the motion-retarding mechanism of the sternum and the costo-sternal joint. It is a lever of the third order, FPW.

2. Of course, motion in but one segment only is impossible. Inspiratory increase in the size of the chest, or *movement* of the *chest as a whole*, is a summation of the movements of single segments. The many fulcra are found at the costo-central joints; the lever is the entire bony chest, including the sternum; the power is supplied at many points on the sides of the chest wall; and the weight and resistance are the weight of the bony chest and its contents plus the resistance of ten pairs of costo-sternal cartilages and the resistance (and complicated) mechanism of the sternum and costo-sternal joints. This compound leverage is of the third order—FPW.

The separate downward motion of the two pairs of floating ribs, in inspiration, uses levers of the third order—FPW.

3. The *complicated mechanism*, due to involvment of the chest with *rotation* of the *upper spine*.

In a previous paragraph (see p. 58), study was made of the simpler mechanism of rotation of the upper thoracic spine alone. This would have been adequate if the movement were not necessarily complicated by the mechanism of the bony thorax. In 1920, Dr. A. Mackenzie Forbes presented a theory of this involved movement, to account for the corrective force he applied to the deformity of Rotary Lateral Curvature of the Spine. Dr. Forbes presented this theory only as applied to corrective plaster of Paris jackets; not to corrective exercises or to what happened to the normal spine and thorax in the movement of rotation. Realizing that this statement of mechanical principles is unproved theory, its application to the writer's Derotation Exercises for Scoliosis is so definite as to warrant kinesiological development as a reasonable statement of the mechanism of the movement in the normal spine and thorax.

Thus, based on Forbes' reasoning of the mechanism, in the deformity of Rotary Lateral Curvature of the Spine, a segment of the chest at the level of the fifth thoracic vertebra is chosen for description (Fig. 17). This segment is divided into four levers—

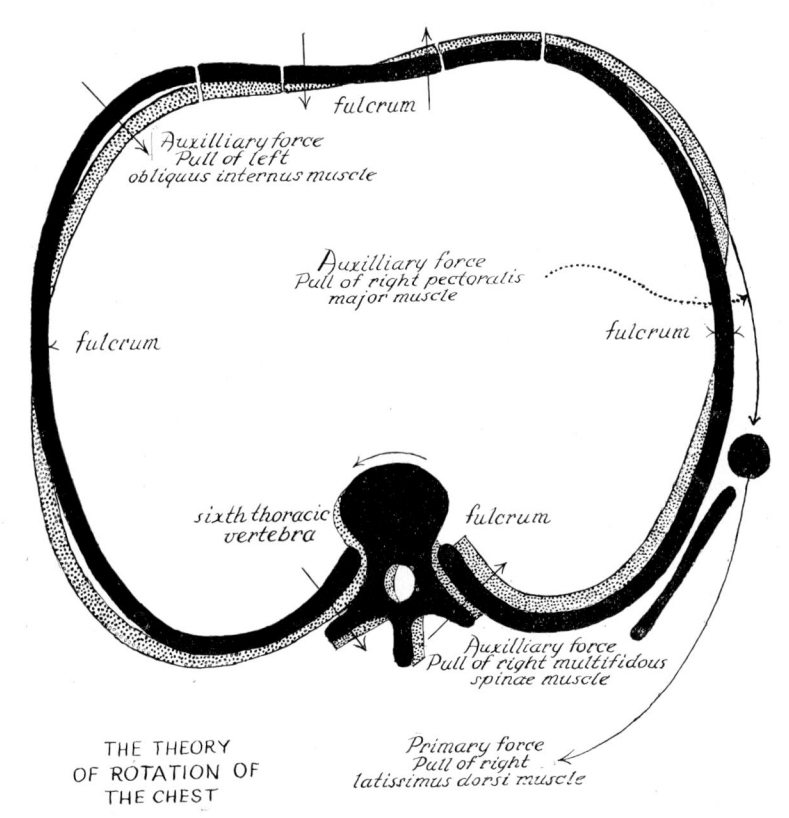

FIG. 17. Illustrating the Theory of Rotation of the spine and thorax, at mid thoracic level, where the upper trunk is gymnastically twisted to the right—that is, the "paradoxical" left twisting of the vertebra with the complemental triple leverage effect on the ribs and the sternum. (For detailed description of the four-leverage mechanism, see text.)

a) the right rib; b) the fifth vertebra; c) the left rib; and d) the sternum. In this compound leverage mechanism, the principal motive factor is that exerted by the strong Latissimus dorsi muscle of the side toward which rotation is made, although other muscles aid. For clearness of statement, upper trunk-twisting to the right will be studied. The Latissimus dorsi of the right side, acting from the pelvis and the lumbar and the lower six thoracic spines, across the rear of the right chest wall and the lower angle of the right scapula and onto the upper end of the humerus, exerts first a bow-string-straightening pressure upon the chest wall and then a forward pulling action (through the humerus, the scapula and the clavicle) on the upper right corner of the sternum. The leverage action of this (to be discussed below) is first upon the right rib, aided by Pectoralis major muscle, then from the right rib upon the thoracic vertebra, then from the vertebra, aided by the Multifidus spinae muscle group, upon the left rib, then from the left rib, aided by abdominal and intercostal muscle groups upon the sternum, and finally from the sternum upon the right rib. More specifically stated,

a) In *leverage* of the *right fifth rib*, the lever is the rib; the fulcrum is assumed to be at about the middle of the shaft of the rib; the power is applied, on its curving rear, as a direct forward pressure of the tightening Latissimus dorsi muscle and by the forward pressed scapula; and the weight (resistance) is that afforded by the transverse process of the vertebra, in the rear. The front of the rib moves forward. This is a lever of the third order, FPW. However, the right Pectoralis major muscle, acting from its humeral insertion on the front chest wall origin, aids the movement of pulling the front of the rib forward, with resistance at the costo-sternal joint. This leverage action of Pectoralis major is on the principle of the third order, FPW, as far as the front of the lever is concerned. But it is to be noted that Pectoralis major's more important function is aid to the Latissimus dorsi in the leverage of the rear of the rib—the forward movement of the rib. This Pectoralis major action involves leverage of the first order, PFW.

b) *Leverage* of the *fifth thoracic vertebra*. The movement is a *left* twist (rotation) of this vertebra upon its fibro-cartilage base— the "paradox" of Forbes. Mechanically it is principally due to the forward drawing of the rear of the right rib which pulls the right transverse process forward (the local ligaments acting as "hauling ropes"). But this left twist of the vertebra is definitely aided by the Multifidus spinae muscle group of the right side, acting from one or more transverse processes below on the spinous process of Thoracic V. Although the entire bone is the lever, the effective part, in this action is the distance from the tip of the right to the tip of the left transverse process, in the case of the right rib action on the transverse process, and that from the body to the tip of the spinous process, in discussing the action of Multifidus spinae muscles. The fulcrum is at the body of the vertebra; the power is applied to the right transverse process and to the spinous process, as noted above; the weight (resistance) is at the left transverse process. The resistance is offered by the head and neck of the rib of the left side. This movement, rotation of the vertebra to the left, thus includes forward motion of the right transverse process and backward motion of the left transverse process. It is a lever of the first order PFW. (Note: The complication of vertebral articular processes, as added factors of resistance, is recognized. But this factor does not change the principle of the movement. Indeed, these articular processes simply add to the effectiveness of this form of leverage as applied to successive levels of the thorax—that is, to the thorax as a whole).

c) *Leverage* of the *left rib*. This bone, which complements and adds muscle-power factors to the combined leverage actions under discussion (and incidentally to the reforming tendency which the combined leverage action gives to the left side of the chest, in the correction of Scoliosis) is important. This fulcrum is also assumed to be at or near the center of the rib. Mechanical power is transmitted from the backward pushing left transverse process to the backward pushed rear end of the rib. In addition, the left Obliquus internus, acting on the left lower front of the chest wall, and transmitting power through successive Internal intercostal muscles upward, furnishes direct muscular action to the front of the left rib. The weight (resistance) is that afforded by the left side of the sternum to the backward drawing front of the rib. This is a lever of the first order, PFW.

d) The *sternum* as a *lever*. This bone may pivot a trifle on its vertical axis, the left side backward and the right side forward.

At any rate, it affords a necessary adjusting factor, made possible by the sterno-costal joint "gives" and by the flexibility of the costal cartilages, to the more definite actions of the three other levers of this quadruple continuous chain.

Application of this principle of compound leverage to the thoracic spine and the bony thorax as a whole, is simply to state that the entire thorax participates, as does the illustrative segment, and that the successive pairs of articular processes of the thoracic spine are important factors in insuring the same action through the entire region.

To recapitulate, this theory of "paradoxical" rotation, in this region, is based on the thesis that there exists a chain of three levers—the right rib, the thoracic vertebra, and the left rib—which exert successive mechanical action, each upon its next in order, and that all have an initiating (Latissimus dorsi) muscle and to each augmenting direct muscle forces are applied; and a fourth lever—the sternum and the costal cartilages—which somewhat eases the strain of the other levers but does not break the chain.

(Note: The writer hopes that other investigators prove, or perhaps authoritatively refute, this theory of the complicated mechanism of rotation of the bony thorax with the thoracic spine.)

Having "built up" the human frame, from the feet to the top of the thoracic spine, in Fundamental Standing Position, there remains the neck and the head of the axial system, and the shoulder girdle, arms, forearms and hands of the appendicular system.

THE NECK—THE CERVICAL SPINE

The cervical spine consists of seven bones (see Figs. 14a and b, p. 46 and Fig. 15, p. 60)—the atlas, the axis and cervical III, IV, V, VI, VII. The five lower cervical vertebrae, III to VII, inclusive, have the general characteristics of the seventeen that have been discussed. In each, there is the body or centrum, separated from neighboring bodies by fibro-cartilage, and a neural arch. The bodies are progressively smaller from below upward. The neural arch of each of these vertebrae is divided into a pair of pedicles, a pair of laminae and a spinous process. Each vertebra has two pairs of articular surfaces and a pair of small transverse processes. Each of the transverse processes has a small passageway for the transmission of important vessels and nerves. The opening, which the neural arch and the body encircles, is larger in the cervical region than below, to accommodate the considerable enlargement of the spinal cord in this region. The axis or second cervical vertebra is similar to the others of the neck, except that there is no fibro-cartilage above its body. Instead, there is definite elongation of its centrum upward, the odontoid (tooth-like) process for special articulation with the atlas. The front of the odontoid process has an articular facet for joining with the front arch of the atlas. (Viewed from the standpoint of homology, this odontoid process, on the axis, might be considered as the body of the atlas, although it is not a part of the atlas.) The *atlas* or first cervical vertebra, has special characteristics. It has no centrum, but instead is supplied with a strong anterior arch, two lateral masses and a neural arch. (The pedicles and the laminae found in other vertebrae are here parts of the anterior and of the neural arches respectively.) The encircled opening, largest in the atlas, transmits the medulla oblongata, an important thickened portion of the spinal cord. The lateral masses have a pair of articular facets below, facing downward, slightly backward and slightly inward to articulate with similar facets on the axis. The upper surface of the lateral masses have a pair of large articular facets to join with the occiput of the skull. The rear center of the anterior arch has a facet for articulation with the odontoid process of the axis.

The Ligaments Attached to the Bones of the Neck

These are the intervertebral fibro-cartilages, the inter-articular cartilages (in pairs), anterior common, posterior common, lateral common (in pairs), capsular (in pairs), ligamentum subflava (in pairs), supraspinous and infraspinous, uniting the cervical vertebrae II to VII, respectively, and cervical VII to thoracic I, each to each; the anterior atlanto-axoidean, the posterior atlanto-axoidean, two capsular, the transverse and the atlanto-odontoid capsular, joining the atlas and axis; the occipito-cervical, the crucial, two lateral odontoid or check and the central odontoid or suspensory, uniting the occiput and axis, and finally the anterior occipito-atlantal, the posterior occipito-atlantal, two capsular and two anterior oblique, joining the atlas to the skull (occipital bone).

The Muscles Attached to the Cervical Spine

These are *origin* of Longus colli, acting on the thoracic spine; Serratus posticus superior, acting on the ribs; Multifidus spinae and Longus colli, acting from lower to higher cervical vertebrae; Obliquus capitus inferior, acting from axis on atlas; Splenius capitus, Tracelo-mastoid, Complexus and Rectus capitus anticus major, acting on the head; Rectus capitus anticus minor, Obliquus capitus superior and Rectus capitus lateralis, acting from atlas on head, and Trapezius, Rhomboids and Levator anguli scapulae, acting on the scapula. The cervical spine gives *insertion* to Erector spinae, Semispinalis dorsi, Semispinalis colli, Longus colli and Splenius colli muscles, acting from below; to Scalenus anticus, Scalenus medius, Scalenus posticus and Cervicales ascendens muscles, acting from the upper chest; to Multifidus spinae and Longus colli, acting from lower to higher cervical spines, and Obliquus capitus inferior muscles, acting from axis to atlas.

THE CERVICO-THORACIC JOINT

This joint, between thoracic vertebra I and cervical VII, is studied separately because 1) the direction of the physiological curves of the spine changes at that point; 2) because the characteristics of the vertebrae involved vary to a greater extent than do those above and those below, each to each; and 3) because this joint marks the junction between the rib-reinforced thoracic spine with the otherwise unsupported cervical spine.

The *Cervico-thoracic* joint is an amphiarthorodial or mixed joint. It is composed of the central articulation, or that between the bodies of the two vertebrae, and a pair of joints which connect the articular processes. The body of Cervical VII rests upon that of Thoracic I, with the interposition of fibro-cartilage. The upper articular surfaces of Thoracic I face upward backward; the lower articular surfaces of Cervical VII face forward downward. The direction of these articular surfaces allows freedom of motion in all directions—flexion, extension, lateral flexion and rotation. The *ligaments* of this joint are practically the same as those found in the thoracico-lumbar joint. Of the *movements*, we note that extension, which is freer than flexion, is limited by the ligaments involved, but especially by the approximation of the component spinous processes each with each. Other motions are limited by the ligaments involved.

THE HEAD

The anatomy of the head will be discussed, in this section, only in-so-far as it relates to the mechanism of Fundamental Standing Position. To detail its structure as the seat of the central nervous system, of the cranial nerves, of the organs of special sense and as the upper passageway for the respiratory and the digestive systems would lead this treatise into fields remote from its purpose.

The *Head* is the top-most segment of the human frame. As such it is divided into the skull and the mandible or lower jaw. The *Skull*, although composed of the many flat bones of the cranium and many very irregular bones composing its base and the face, is immovable in its parts, each to each. The *Mandible* will herein be dismissed after stating that it is a movable part, held to the skull by ligaments and controlled by muscles, but that this separate control is so definite and complete as to be disregarded in studies of posture. The head, as a body segment, will therefore be our consideration.

The head is rounded, somewhat ovoid, in its upper rear two-thirds, and comparatively flattened in its lower front one-third. The "point of the chin" is a prominent feature. Four of its twenty bones, the occiput, the two temporals with their mastoid processes, on the lower and lower lateral aspects of the skull, and the mandible, are specially named, because to them are attached important ligaments (to the occiput) and muscles (to all four) which affect posture.

The *Occiput* is a flat bone at the base of the skull, having a cerebral or internal concave surface and an external convex surface. At the rear of the external surface, is an elevation known as the inion or external occipital protruberance. Below this, are three pairs of transverse ridges known as nuchal lines—the superior, the middle and the inferior. This protruberance and these ridges and the spaces between them give attachment to important ligaments and muscles. Near the front of the bone, is a large opening, known as the foramen magnum, through which passes the medulla oblongata. At the sides of this opening and near the front, are two rounded rather massive condyles, which rest on and articulate with the lateral masses of the atlas. Just external to these condyles are a pair of jugular processes, which afford attachment to ligaments and muscles, and in front of the foramen magnum is the basi-occipital process, to which are also attached ligaments and muscles.

The *Temporal* bones are a pair of partly flat and partly irregular bones situated at the side of the skull and affording attachments to ligaments and muscles. One-half of the inner surface of each is concave, while the other half is the petrous portion, a very irregular inward backward projection of bone, through which passes important vessels and nerves. The outer surface is partly smooth and convex and partly very irregular. The most prominent feature of this irregular portion is a deep canal, known as the auditory meatus, connecting the external ear with the latter's internal mechanism. To the rear of this, is a large irregular mass, known as the mastoid process, which affords attachment to muscles. Finally, extending downward and slightly forward from the meatus, is a delicate projection, known as the styloid process, for muscle attachment.

The *Mandible* or lower jaw, situated at the lower front of the head, is like a horse-shoe in shape, extending from the prominent mental protruberance or typical feature of the jaw in front, backward outward, on each side, to the angles of the jaw, and then upward to a coronoid process, in front, and to a condyle behind. The curved portions on each side are known as the bodies and afford attachment to muscles. The upper portion of the semi-circular body has a series of depressions for the lodgment of the lower teeth. The upward extending portions on each side are comparatively flat but massive. These are called the rami and afford attachment to muscles. Each ramus is nearly quadrilateral in shape, except that the upper border is concave to the extent of nearly half a circle. The coronoid process affords attachment to muscles. At the upper rear of each ramus, is a neck and a condyle, for articulation with the temporal bone and for the attachment of ligaments. The ligaments attached to the mandible are entirely for control of its joints with the skull. Most of the muscles attached to the mandible also regulate the movements of the jaw on the skull. For the purpose of this discussion, it is sufficient to state that, in the Fundamental Standing Position, the ligaments and muscles so completely control the jaw that we may consider the head as one piece and will not go further into that mechanism. On the other hand, the mandible affords attachment to some muscles extending from the human frame below which are important in the consideration of posture.

Ligaments Attached to the Head (Not Including Intrinsic Ligaments or Those Controlling the Lower Jaw)

These are the occipital-cervical, the crucial, two lateral odontoid or check and the central odontoid or suspensory, connecting the axis and the occiput; and the anterior occipito-atlantal, posterior occipito-atlantal, two capsular and two anterior oblique, connecting the atlas and the occiput.

Muscles Attached to the Head

The head affords *origin* to the Trapezius, which acts from it (and from all of the cervical and thoracic spine) on the scapula. The head gives *insertion* to Splenius capitus, Trachelo-mastoid, Complexus and Sterno-mastoid, acting from the thoracic spine, the ribs or the shoulder girdle; to the Rectus capitus anticus major, acting from the cervical spine, and to the small but important Obliquus capitus superior, Rectus capitus lateralis and Rectus capitus anticus minor, acting from the atlas.

JOINTS OF THE NECK

The joints of the neck are divided into those between 1) the six lower, the second to seventh, inclusive; 2) that between the axis and the atlas; 3) that between the atlas and the occiput; and 4) that between the axis and the occiput.

1. THE CERVICAL JOINTS.

These do not differ much from the cervico-thoracic joint which has been described. They are amphiarthrodial, or mixed joints. There are fibro-cartilages, between the bodies of consecutive vertebrae, and the articular cartilages between the articular surfaces. The anterior common ligament, the posterior common ligament and the lateral common ligaments (in pairs) connect the bodies; the capsular ligaments (in pairs) connect the articular surfaces, the ligamenta subflavae (in pairs), connecting the the laminae, and the supraspinous and the infraspinous ligaments which connect the spinous processes. The articular surfaces of the articular processes face respectively upward backward (for the upper facets) and downward forward (for the lower facets), allowing all motions—flexion, extension, lateral motions, rotation and circumduction. It is to be noted that extension is freer than flexion and freer here than is extension in the lumbar region, that flexion is somewhat more limited in the neck than in the lumbar region and that rotary motions are free in the entire neck and especially in the lower half. (The kinesiology will be discussed later—see p. 84.)

2. THE ATLANTO-AXOIDEAN JOINTS.

These are diarthrodial joints and are divided into those between the upper articular processes of the axis and the lower surfaces of the lateral masses of the atlas and that between the front of the odontoid process of the axis and the anterior arch of the atlas. Although some flexion, extension and lateral motion are present in these joints, the principal motion is rotation. This consists in a circular motion to the right or left of the atlas (carrying the head) around the fixed odontoid process of the axis. In this motion, if there is head twisting to the right, for instance, the right lateral mass of atlas slides backward on the upper articular surface of the axis, and the left lateral mass of the atlas slides forward on the left side of the axis. As we would expect, to regulate these movements there are special ligaments. They are the anterior atlanto-axoidean, the posterior atlanto-axoidean, two capsular (for the lateral joints), controlling the sliding articulations referred to, and the Atlanto-odontoid capsular ligament and the transverse ligament, regulating the pivoting action of the atlas around the odontoid process.

3. THE OCCIPITO-ATLANTAL JOINT.

The occipital bone has two rather massive processes, known as condyles (see p. 77), on either side of the foramen magnum, projecting downward, for articulation with the lateral masses on the upper surface of the atlas. The joint therefore allows movement between these condyles and the articular surfaces of the atlas. The motion in these joints is not great, simple head flexion (nodding) and extension being the principal ones. But a certain amount of lateral motion is possible. Because of the hinge-like effect, this has been classed as a ginglymous joint. This limited action between the head and atlas is necessitated for the protection of the medulla oblongata, which at this point extends from the lower end of the brain to the top of the spinal cord. Were motion too free at this joint, there would be constant danger of harmful pressures of the edge of the foramen magnum on the medulla. To insure limited action, the ligaments of this joint and of the occipito-odontoid joint (see below) are comparatively strong and so situated as to check excessive motion. The *ligaments* of this occipito-atlantal joint are anterior occipito-atlantal, posterior occipito-atlantal, two capsular and two anterior oblique.

4. THE OCCIPITO-ODONTOID JOINT.

There are no articular surfaces between the occiput and the odontoid process, but there are strong ligaments extending between these bones, the main function of which is to limit undue, harmful motion. These *ligaments* are occipito-cervical, the crucial, two lateral odontoid or check and the central odontoid or suspensory.

MOVEMENTS AND MUSCLES OF THE NECK AND HEAD

These will be divided into: 1) Those of the neck (exclusive of atlas and head); 2) Those between axis and atlas; 3) Those between atlas and head; and 4) Those of the entire segment—neck and head.

1. In the *cervical region* are found flexion, extension, lateral flexion, circumduction and rotation.

a) *Flexion* is fairly free. In standing position, it is *caused* by gravity and controlled by extensor muscles, acting eccentrically. But it is initiated by Scalenus anticus and Longus colli muscles, acting from the first rib, from upper thoracic and lower cervical vertebrae on higher cervical vertebrae. If the body is recumbent on the back, flexion is due to Sterno-mastoid muscles, acting from the upper thorax on the neck, by Scalenus anticus, Scalenus medius, Scalenus posticus and Longus colli muscles, acting on the thoracic spine, and by Rectus capitus major, acting from the cervical spine on the head. Flexion is *limited* by the approximation of the chin on the chest and by the interspinous and supraspinous ligaments.

b) *Extension* is free in the neck. Although gravity and the regulating flexor muscles play their part, in standing position, the intrinsic muscles for extension are Erector spinae, Semispinalis dorsi, Multifidus spinae, Splenius colli, Semispinalis colli, Complexus, Splenius capitus and Trachelomastoid, Trapezius (upper portion) and Levator anguli scapulae, all acting from the pelvis, the lower spine, the chest or the shoulder girdle. Extension is *limited* by the anterior common ligaments and by the approximation of the spinous processes.

c) *Lateral Flexion* is very free in the neck and is *caused* by Erector spinae, Semispinalis dorsi, Multifidus spinae, Splenius colli, Scalenus anticus, Scalenus medius, Scalenus posticus and Trapezius, acting from the pelvis, the thoracic spine, the ribs and the shoulder girdle on the neck; by Longus colli, acting from cervical spine to cervical spine, and by Splenius capitus, Trachelomastoid and Complexus, acting from the thoracic spine on the head. These movements are initiated by the muscles of the side toward which motion is made. Lateral flexion is *limited* by the lateral ligaments of the side opposite to the motion.

d) *Circumduction* is *caused* by the consecutive action of flexion, lateral motion, extension and opposite lateral motion of the head and neck and is *limited* by the factors limiting the component movements. (Rotation is not a part of true circumduction. It is to be noted that flexion and extension are caused by bilateral action of the muscles involved, lateral flexion and rotation by unilateral muscular action and circumduction by serial action of the muscles of flexion, lateral motion, extension and opposite lateral motion.)

e) *Rotation* is fairly free in the cervical region (although head turning is also produced by free rotation between axis and axis— (see below, p. 84). Rotation is *caused* by Splenius colli and Longus colli muscles, acting from the thoracic vertebrae to twist the cervical vertebrae to the same side; by Semispinalis dorsi, Semispinalis colli and by Multifidus spinae muscles, acting from the thoracic vertebrae to turn the neck to the opposite side; by Rectus capitis anticus major, Splenius capitis and Trochelo-mastoid muscles, acting from upper thoracic and lower cervical spine on the head, to turn neck to the same side; by Complexus, acting from thoracic and cervical spine on the head, to twist the neck to the opposite side; by Scalenus anticus muscle, acting from the ribs on the cervical vertebrae, to rotate the neck to the opposite side; and by the powerful Sterno-mastoid, acting from the sternum and clavicle on the head to turn the neck to the opposite side. (Note: In torticollis or wry neck, the Sterno-mastoid of one side is contracted or shortened causing the head to be bent to the same side and the face to be turned to the side opposite to the muscle involved.) Rotation in the neck is *limited* by ligaments between vertebrae, especially by the capsular ligaments.

2. In the *Joints between Axis and Atlas*, we have principally rotation, although slight flexion, extension and lateral flexion is possible. (The musculature of flexion, extension and lateral flexion are covered in the paragraph on the cervical joints.)

Rotation of atlas on axis is *caused* by many of the same muscles as those for rotation of the neck, and in addition by Obliquus capitis inferior, acting directly from the axis on the atlas to turn the head and atlas to the same side. Rotation between these two bones is *limited* by the check ligaments and the atlanto-axoidean ligaments.

3. The *motions* between the *atlas* and the *head* (*occiput*) are

flexion, extension and slight amount of lateral motion. There is no rotation.

a) *Flexion*, in the standing position, is *caused* by gravity, controlled by extensor muscles, but is initiated by Sterno-mastoid, acting from the bony thorax on the head; by Rectus capitus anticus major, acting from the cervical vertebrae on the head, and by Rectus capitus anticus major, acting from the atlas on the occiput. Flexion between these two segments is *limited* by capsular and cervico-occipital ligaments.

b) *Extension* is *caused* by Obliquus capitus superior muscle, acting from atlas to occiput; by Splenius capitus, Trachelo-mastoid and Complexus, acting from the thoracic and cervical spines on the head, and by Trapezius (upper portion), acting from the shoulder girdle on the head. All of the muscles act in pairs to cause extension of the head on the atlas. Extension is *limited* by the anterior oblique and the anterior occipito-atlantal ligaments.

c) *Lateral flexion* is very slight and is a combination of the action of the flexors and the extensors of the same side, and also by Rectus capitus lateralis muscle, acting from atlas on the occiput. Lateral flexion is *limited* by lateral odontoid ligaments and by the capsular ligament of the opposite side.

4. MOVEMENTS AND KINESIOLOGY OF THE HEAD AND NECK AS A WHOLE.

The carriage of this compound segment is an important factor in posture. A drooping head or a laterally carried head affects the line of gravity and disturbs body balance. Backward tilt of the head is rarely a feature to be corrected. The parts of the head, carried in front of the condyloid bases of the occiput, are heavier than those to the rear. Because of this, the muscles which extend the head and neck are definitely stronger than the flexors. The movements in this region are those of the component parts, already enumerated, and are controlled by the muscles which have been named. The muscles of this region must play a more important part than in the thoracic and lumbar spines, because the cervical spine is much less massive and less provided with bony stays than in the mid and lower spine. In ordinary activities, the neck and the head motions are in the same direction—that is, flexion of the head on the neck accompanies flexion of the neck, as also do extension and rotation. But flexion and extension of the head on the upper spine may be done apart from the same motions of the neck alone. However, in faulty posture, with forward drooping head and neck, correction is not made by the same directional movement. If it were, extension of the neck would be accompanied by backward tilting of the head. This correction, therefore, consists in extension of the neck and slight though definite flexion of the head on the atlas. This is a procedure often difficult to train, but is important. Rotation of the neck and of the head on the neck (atlas on axis) usually accompany each other; but either alone may be possible. In the gymnastic movement of head twisting to the right or to the left, for instance, the untrained person usually flexes head and neck (in both segments). To do this movement correctly, instructions must include training to keep the neck extended and the head slightly flexed (see above) while making the head-neck twist. This is necessary to avoid carrying this segment out of the line of gravity.

Leverage in this region follows the principles previously stated.

THE UPPER EXTREMITY

Introductory: In the study of Fundamental Standing Position, the human frame has been "built-up" from a bilateral pedestal—feet, legs, thighs—, through a unilateral body—pelvis, lumbar spine, thoracic spine, cervical spine and head—, and including a bilateral though closely axial, chest wall. There remains the distinctly bilateral appendages, known as the upper extremities. These, in standing position, are added weights and not added supports to the human structure.

The upper extremity is divided, for anatomical description, into the shoulder girdle, the arm, the forearm, the wrist, the hand and the fingers; and includes the sterno-clavicular, the acromio-clavicular, the shoulder, the elbow, the radio-ulnar, the wrist, the metacarpo-phlangeal and the phlangeal joints and finally the muscles controlling these segments.

THE SHOULDER GIRDLE

The *Shoulder Girdle* includes the clavicle and the scapula (Fig. 18a). The clavicle joins the upper outer corner of the

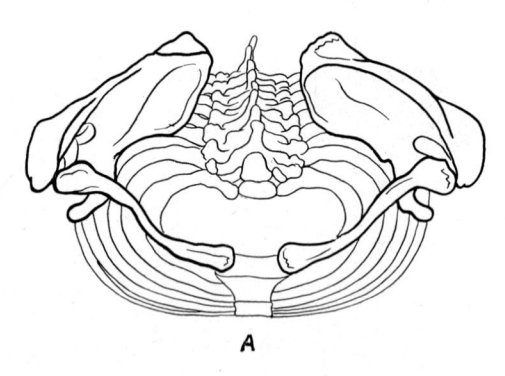

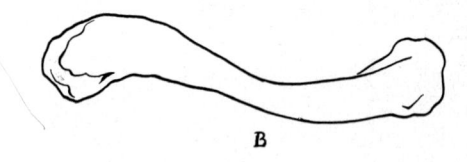

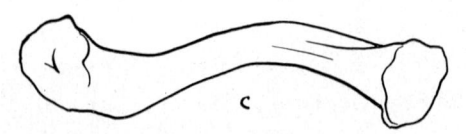

FIG. 18. A. Shoulder girdle, from above, $\frac{1}{10}$ life.
B. Right clavicle, from above, $\frac{1}{2}$ life.
C. Right clavicle, from below, $\frac{1}{2}$ life.

sternum at its inner end; the scapula articulates with the outer end of the clavicle at its (the scapula's) acromiun process, and the scapula gives joint union, at its glenoid cavity, with the head of the humerus or upper arm bone.

1. The *Clavicle* (Figs. 18b and c) is a long somewhat flattened bone, having two curves, an inner, with convexity forward, and an outer, with convexity backward. It also has an upper and lower surface and two articular ends. It is comparatively thin in the middle (the most common location for fraction of this bone) and has somewhat broadened extremities, especially the outer. The clavicle affords attachment to four sterno-clavicular *ligaments*—capsular, interclavicular, interarticular fibrocartilage and rhomboid or costo-clavicular, at the inner end of the bone, and attachment to four sterno-clavicular ligaments—a capsular, interarticular, conoid and trapezoid—at its outer end. The clavicle gives *origin* to Sterno-mastoid muscle, acting on the head and to Pectoralis major and Deltoid muscles, acting on the humerus. This bone gives *insertion* to Trapezius, acting from the head, the cervical and the thoracic spine, and to Subclavius, acting from the ribs.

2. The *Scapula* (Figs. 19a and b) is a large flat bone, triangular in shape, situated near the upper outer rear of the thorax (to which, however, it has no joint attachment). It has two surfaces, an anterior and a posterior; three borders, an upper, an outer or axillary, and an inner or vertebral, and three angles, an anterior, a posterior-superior and a posterior-inferior. The *posterior surface* is divided into two unequal portions or fossae—the smaller or upper and the larger or lower—by a prominent bony process known as the spine of the scapula. This spine of scapula extends horizontally from the vertebral border, increasing in height and bulk as it passes outward and slightly upward to terminate in a broadened acromium process. The large fossae afford attachment to important muscles, the Supraspinatus above and the Infraspinatus below. The acromium process gives *origin* to the Deltoid and the short head of the Biceps muscles. It also affords *insertion* to the Trapezius and attachment to coraco-acromium ligament, whose purpose is to afford extra protection to the head of the humerus. The *anterior surface* is comparatively flat throughout and affords *origin* to the Subscapularis. The *upper border* is situated at about the level of the upper curve of the thorax. Near its outer end is a small groove for the transmission of a muscle and an artery and it is protected by a small transverse

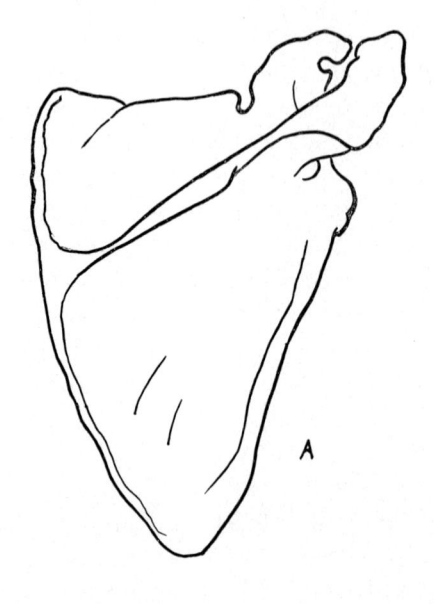

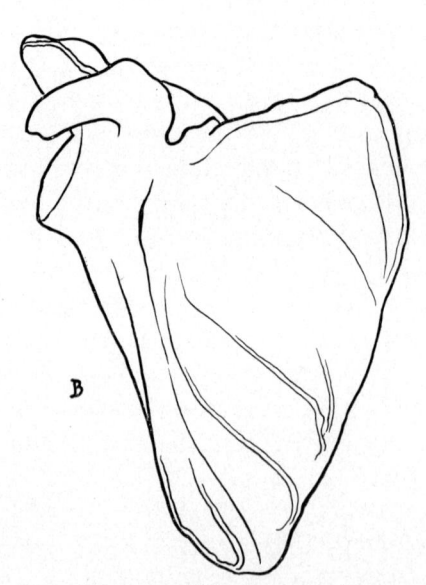

FIG. 19. Right scapula, ⅓ life.
A. Rear view.
B. Front view.

ligament. The *outer* or *axillary border* of the scapula takes a general direction downward and somewhat inward from its upper end and forms the rear boundary of the armpit or axillary cavity. This border gives *origin* to the long head of the Triceps muscle near the shoulder joint and to the Teres minor muscle. The *inner* or *ventral border* is the longest edge of the scapula and lies in an almost perpendicular position. It gives *insertion* to the Serratus magnus and the Rhomboid muscles. The *Anterior angle* of the scapula is a thickened and nearly rounded process, the principal function of which is its development into a shallow saucer-like depression, known as the glenoid cavity, for articulation with the head of the humerus. It gives attachment to the capsular ligament of the shoulder joint, the glenoid-humeral, coraco-humeral and the glenoid fibro-cartilage. The anterior angle also has a strong curved extension known as the coracoid process, extending from near the upper margin of the glenoid process upward and outward. It also affords added protection to the shoulder joint and gives attachment to the acromio-coracoid ligament. Indeed, the outer end of the acromium process, the coracoid process and this strong acromio-coracoid ligament afford considerable protection to the shoulder joint. The coracoid process also gives *origin* to the short head of the Biceps and to the Coraco-brachialis muscles and *insertion* to the Pectoralis minor. The *posterior-superior angle* of the scapula forms nearly a right angle. It is rather thicker and more massive than its neighboring upper and ventral borders. It affords *insertion* to the Levator anguli scapulae and to Serratus magnus muscles. The *posterior-inferior angle* is more acute than either of the others. It is also thickened and affords *origin* to the Teres major and occasional origin to the Latissimus dorsi and *insertion* to the Serratus magnus muscles.

THE HUMERUS

The Humerus or upper arm bone is situated between the scapula and the ulna (Fig. 20a). It is a long bone, having an upper extremity, a shaft and a lower extremity. The upper extremity has a head, a neck and two tuberosities. The head has a rounded

90   BODY POISE

articular surface, described as one-third of a sphere, for articulation with the glenoid cavity of the scapula. This articular surface is covered with fibro-cartilage and faces upward inward. The neck is very short, but gives attachment to the capsular ligament

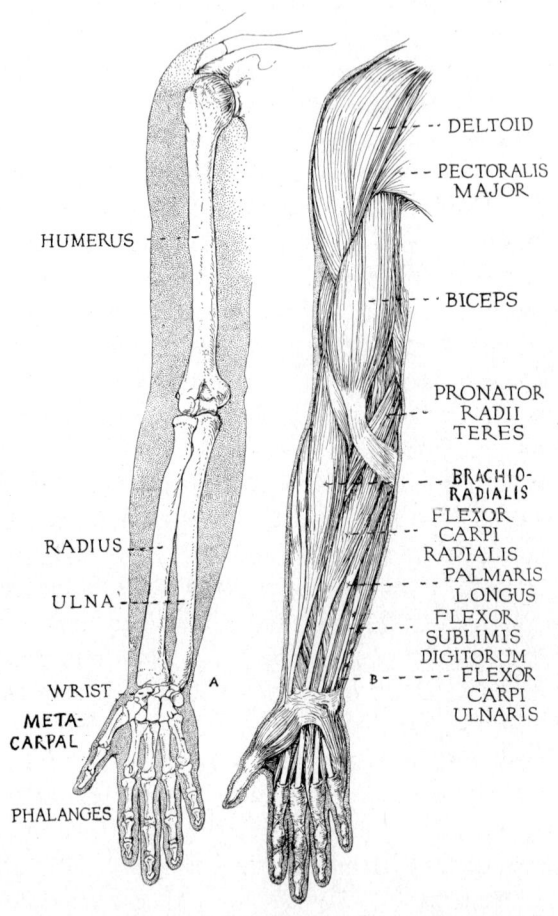

FIG. 20. A. Bones of the right upper extremity, front view, $\frac{1}{6}$ life.
B. Muscles of the right upper extremity, front view, $\frac{1}{6}$ life.

and insertion to Subscapularis muscle in front. The tuberosities are just external to and below the neck. The greater tuberosity above, gives attachment to Supraspinatus and Infraspinatus muscles. The lesser tuberosity at the rear of and below the greater tuberosity, gives attachment to the gleno-humeral ligaments and insertion to the Teres minor muscle.

Just below the upper extremity of the humerus is a somewhat narrowed portion of the shaft, which is called the surgical neck, because the upper arm bone fractures are more liable to occur at that point than elsewhere.

The shaft of the humerus is cylindrical in its upper third, cylindrical in its middle third and rather flattened laterally in its lower third. It has certain bony lines and small elevations for the attachment of muscles. The upper third gives *origin* to the external head of Triceps muscles, in the rear, and *insertion* to Latissimus dorsi, Pectoralis major and Teres major in front. The middle third gives *origin* to the internal head of Triceps muscle in the rear and *insertion* to Deltoid at the outer side and Coraco brachialis, at the inner side. The lower third gives *origin* to Brachialis anticus muscle, in the front, to Supinator longus and Extensor carpi radialis longior at the outer side and to the internal head of Triceps at the rear.

The lower extremity of the humerus has an external and an internal condyle, a coronoid fossa for the reception of the coronoid process of the ulna, in front, an olecranon fossa, in the rear, for the reception of the olecranon process of the ulna and an irregular shaped articular surface for articulation with the forearm bones. The external condyle gives *origin* to Extensor carpi radialis brevior, Extensor communis digitorum, Extensor minimi digiti, Extensor carpi ulnaris, Supinator brevis and Flexor carpi ulnaris muscles. The inner condyle gives attachment to external lateral ligaments and *origin* to Pronator radii teres, Flexor carpi radialis, Palmaris longus, Flexor sublimis digitorum, Flexor carpi ulnaris and Anconeus. (This bone is a good example of the general principle of muscle insertion attachments at or near the upper or more fixed portion and of original attachments at or near the lower or more movable portion.) The articular surface at the lower end of the humerus is quite uneven. From side to side it is an irregular cylinder, with curving surfaces from before backward constituting a nearly complete circumference, the deep fossae just above leaving but a thin plate of bone. However, these articular surfaces differ, from side to side; on the inner side the trochlea is saddle-shaped, facilitating a true hinge joint with the ulna. At the outer side, the capitellum presents a ball-like knob for the double action upon it of the head of the radius—a hinge action and a fixed point for separate pivot-like rotation of the head of the radius. Completely encircling the entire articular surface at the lower end of the humerus, there is attachment of the capsular ligament.

(In former times, the name *cubit* was given to this bone, the humerus. Among the ancients and in Biblical times, the cubit was also a measure of length. It is said that, among the Egyptians, the measured length of the reigning Pharaoh's upper arm bone was taken as the standard. One wonders at the disturbance in trade and commerce which might follow the death of a king.)

THE JOINTS OF THE SHOULDER GIRDLE

The *Joints* of the *Shoulder Girdle* are the sterno-clavicular and the scapulo-(or acromio-) clavicular joints.

1. THE STERNO-CLAVICULAR JOINT.

This is an arthrodial joint, connecting the inner end of the clavicle with the upper outer corner of the sternum. Between these two bones is an articular fibro-cartilage, which gives freedom of motion. The ligaments governing the joint are capsular, interclavicular, interarticular and costo-clavicular. Motions in this joint are upward, downward, forward, backward and circumduction. The ligaments, especially the capsular and the rhomboid, which limits upward motion, allow but slight degree of motion in any direction. There are no rotations at this joint.

a. *Upward* motion is *caused* by Trapezius (upper portion) and Sterno-mastoid muscles (reverse action), acting from head on clavicle, and by Levator anguli scapulae and the Rhomboid muscles, acting from cervical spine on scapula. Upward motion in *limited* by the sterno-clavicular ligaments and by the rhomboid ligament.

b. *Depression* of the clavicle is *caused* by Latissimus dorsi muscle, acting from the pelvis on the humerus, by Pectoralis minor, acting from ribs on scapula and by Subclavius, acting from ribs on clavicle. Depression is *limited* by the approximation of the clavicle on the upper thorax.

c. *Forward motion* of clavicle is *caused* by Pectoralis major muscle, acting from chest on humerus, by Pectoralis minor, acting from the chest on scapula and by Subclavius, acting from the chest on clavicle. Forward motion is *limited* by the sterno-clavicular ligament.

d. *Backward motion* of clavicle is *caused* by Latissimus dorsi muscle, acting from pelvis on humerus, and by Trapezius and Rhomboids, acting from the spine on the scapula. Backward motion is *limited* by sterno-clavicular and the rhomboid ligaments and by the approximation of the clavicle on the upper chest wall.

e. *Circumduction* is a consecutive movement of elevation, forward motion, depression and backward motion—or a reversal of this—and is *caused* by and *limited* by the muscles and ligaments, in series, that cause and limit the elements of the component movements.

2. THE ACROMIO-CLAVICULAR JOINT.

This is an arthrodial joint, connecting the outer end of the clavicle with the acromium process of the scapula. There is an interarticular fibro-cartilage, to facilitate freedom of movement, and a capsular ligament to limit movement and dislocation. The movements at the acromio-clavicular joint are those reciprocating with the movements at the sterno-clavicular joint (a and b below) and those in which the scapula adds scope to the general direction of the movements of the sterno-clavicular joint (c, d, e, f below).

The *scapular movements*, which reciprocate with clavicular movements, will be understood by describing what takes place at the tip of the shoulder—thus:

a. In true *elevation* of the shoulder, there is elevation at the sterno-clavicular joint and a corresponding inward movement of the scapula at the acromio-clavicular joint to keep the arms in the neutral position, at the sides of the body. This is shrugging of the shoulders.

b. In *depression* of the shoulder, there is depression at the sterno-clavicular joint and there is a corresponding returning outward movement of the scapula, at the acromio-clavicular joint, to keep the arms in the neutral position.

In both of these motions, the movements are caused by and limited by those described above in discussing the movements at the sterno-clavicular joint.

In forward and in backward motion of the shoulder girdle, there is usually an added excursion of the scapula, centered at the sterno-clavicular joint, increasing the total movement. Thus:

c. In *forward* movement, the scapula abducts and, although the Pectoralis major muscle is the principal cause of the forward movement of the girdle, there is "winging" of the scapula, by action of the Pectoralis minor, acting from the upper ribs on the coracoid process of the scapula.

(This is the position, which, if it becomes habitual, is part of so-called round shoulders, usually due to weakness of the shoulder adductors, the Trapezii especially. On the other hand, it may be present in those who over-train the Pectoralis major muscles as in parallel bar work in the gymnasium and in the posture of prize fighters, if they do not train against it.)

d. In *backward* movement of the shoulder girdle, there is backward motion of the clavicle plus scapular adduction. (This is the position of good military posture.)

There are two mutually directional sterno-clavicular and acromio-clavicular motions. They are elevation of the clavicle with upward rotation of the scapula and depression of the clavicle with downward rotation of the scapula. The movements are best understood by noting that the center of motion, the pivot, is at the sterno-clavicular joint and that the added movement at the glenoid cavity, up or down, expresses the rotation of the scapula upward or downward. In pure rotations of the scapula, the bone maintains its parallel relationship to the outer upper rear of the chest wall, and it is seldom taken without abduction and adduction of the humerus.

Elevation and depression of the clavicle have been described (p. 92).

e. *Upward rotation* of the scapula is *caused* by Trapezius muscle, acting from spine on the scapula, and Serratus magnus, acting from the chest wall on scapula. Rotation upward is very free but is finally *limited* by the tension of the capsular ligament at the acromio-clavicular joint.

(It will be noted below (p. 112), that the gymnastic movement of arm raising sideways upward is a composite one of abduction of the humerus at the shoulder joint, upward rotation of the scapula and elevation of the outer end of the clavicle.)

f. *Downward rotation* of the scapula is *caused* by Latissimus dorsi muscle, acting from the pelvis on humerus; by Rhomboids, acting from thoracic spine on scapula; by Pectoralis major, acting from the chest wall on scapula and by Levator anguli scapulae, acting from cervical spine on scapula. Downward rotation of the scapula is *limited* by the tightening of the capsular ligament of the acromio-calvicular joint.

(This downward rotation of the scapula and the muscles which control it are important factors in climbing, in swimming and in some of the corrective exercises to be discussed in Part II.)

KINESIOLOGY OF THE SHOULDER GIRDLE

Emphasizing what has been stated in the Anatomy of this region, it is to be pointed out that the maintence of adducted scapulae, or laterally straight upper back, is essential in good position. The converse of this, or round shoulders, is often a concomitant of flat feet, relaxed knees, hollow back with pendulous abdomen, round back and sagging head and neck. Burroughs wrote "I note that when a family begins to run out (sic), it protrudes at the waist, flattens the chest, recedes at the chin and dotes on small hands and feet."

THE SHOULDER JOINT

The shoulder joint is an enarthrodial or ball-and-socket joint, uniting the glenoid cavity of the scapula with the head of the humerus. The glenoid cavity is quite shallow and lined with articular cartilage and its rim gives attachment to an added cartilagenous structure, which deepens the joint cavity. The articular surface on the head of the humerus is also covered with cartilage. The bones are held together by ligaments and muscles.

The Ligaments of the Shoulder Joint

They are capsular and coraco-humeral.

a. The capsular ligament is a loose sac, attached around the rim of the glenoid cavity, on the one hand, and to the edges of the articular surfaces of the head of the humerus, on the other. It limits humeral movements in all directions.

b. The coraco-humeral ligament is a strong fibrous band which reinforces the upper portion of the capsular ligament. It is attached to the coracoid process above, blends with upper fibres of the capsule and is again attached to the greater tuberosity of the humerus.

These two ligaments are strong but insufficient to maintain the head of the humerus from dislocation in severe trauma. Tendons of muscles passing near the joint are essential in preserving the integrity of the joint. Indeed, in some paralyses of the muscles involved, there is an habitual partial dislocation, usually with the head of the humerus forward. On the other hand, the shoulder joint is protected from violence from above by the acromium and coracoid processes on the scapula and the strong coraco-acromium ligament connecting these processes.

The Movements and Muscles of the Shoulder Joint

The movements of the shoulder joint are a) flexion; b) extension; c) abductions; d) adduction; e) circumduction; f) rotation inward; and g) rotation outward.

a. *Flexion* is *caused* by Pectoralis major muscle, acting from the chest; by Deltoid (anterior portion), acting from the clavicle, and by Biceps and Coraco-brachialis, acting feebly from the scapula. Pure flexion at the shoulder joint is *limited* by the approximation of the neck of the humerus on the coracoid process of the scapula, at arm position forward of about ninety degrees; but from this point upward rotation and abduction of the scapula allow arm carriage upward to the perpendicular.

b. *Extension* is *caused* by Latissimus dorsi muscle, acting from the pelvis and by Deltoid (rear portion), acting from the spine of the scapula. Pure extension at the shoulder joint is *limited* by tightening of the capsular ligament, but adduction and downward rotation of the scapula allow further backward carriage of the arm to about forty-five degrees from the lateral plane of the body.

c. *Abduction* at the shoulder joint is *caused* by Deltoid muscle, entire, acting from clavicle and scapula and by Supraspinatus, acting from the scapula. Pure abduction at the shoulder joint is *limited* by the approximation of the neck of the humerus on the acromium process. (The interesting relationship of shoulder joint and shoulder girdle will be discussed under Kinesiology—see p. 111.)

d. Pure *adduction*, in this joint, in Fundamental Standing Position, is impossible as the forearm is already in approximation with the side of the pelvis. Adduction-flexion and adduction-extension will therefore be discussed.

Adduction-flexion is *caused* by Pectoralis major muscle, acting from the chest and by Biceps and Coraco-brachialis, acting from scapula. All adduction-flexion is accompanied by forward motion of the clavicle and adduction of the scapula and the movement is *limited*, at about forty-five degrees, by the tightening of the rhomboid ligament between the upper ribs and the clavicle, and approximation of upper arm on the front-mid-chest.

Adduction-extention is *caused* by Latissimus dorsi, acting from the pelvis and by Infraspinatus, Teres major, Teres minor, Subscapular and Triceps, acting from the scapula. This movement is associated throughout by backward motion of the clavicle and by adduction of the scapula. It is *limited*, at about thirty degrees, by the approximation of the clavicle on the upper ribs, by the approximation of the scapula to the rear mid-chest wall and by tightening of the capsular ligament of the shoulder joint.

e. *Circumduction* is a consecutive motion of flexion, abduction, extension and adduction (or the reverse) with motive power and limitations as named in the integral factors. Circumduction may be considered as the formation of an imaginary cone, the apex of which is the shoulder joint, the base the circle formed by the finger tips and the sides by the moving arm, forearm, hand and fingers. As a gymnastic exercise, the full limits of the movement are not made; that is, adduction does not include reaching the plane of the side of the body (especially at the rear horizontal). Also the imaginary cone has real, though small, altitude but a large base, in the gymnastic movement sometimes called the "windmill."

The rotations, inward and outward, will be discussed as in the anatomical position of the arm at the side of the body, although rotations may be made in any angular variation from this position, which variations are often used in gymnastic drills.

f. *Rotation inward* is *caused* by Latissimus dorsi muscle, acting from the pelvis, by Pectoralis major, acting from the chest wall, by Subscapularis, by Teres major and by Deltoid, front portion, acting from the scapula. As each of these muscles is inserted into the humerus at a site partly curved around the upper humeral shaft, the *limit* of internal rotation is reached when the muscle or its tendon is entirely uncoiled and the pull of the muscle is thus directly away from the bone. Pure internal rotation is possible for about ninety degrees.

(Note: An added intwist of the entire upper extremity is caused, not at the shoulder joint, but by pronation of the forearm. To eliminate forearm intwist or out-twist, as a gymnastic exercise a muscularly fixed right-angled elbow is used.)

g. *Rotation outward* is *caused* by Infraspinatus, Teres minor and Deltoid, rear portion, acting from the scapula. This is *limited* by the tortion tenseness of the capsular ligament of the shoulder joint. Pure outward rotation at the shoulder joint is possible for about sixty degrees. (Here, too, supination of the forearm may add a little to the total out-twist of the entire upper extremity.)

THE FOREARM

Two bones, the ulna and the radius, comprise the skeletal part of the forearm (see Fig. 20a, p. 90). They are united with the humerus at the elbow joint, with each other at the upper and lower radio-ulnar joints, and below they are continuous with the carpus or wrist bones. The bellying of muscles produces a greater circumference, in the upper two-fifths, as the preponderance of tendons, in the lower three-fifths of the forearm, considerably lessens the girth. In the bellied part, the girth is almost a complete circle; in the tendinous section, the general form is somewhat flattened with rounded sides.

ULNA

This is a long bone, having an upper extremity, a shaft and a lower extremity. It is on the inner side of the forearm. The upper extremity is irregular in shape. At the top rear is the olecranon process, which curves slightly forward at its tip. Near the top, in front of the bone, is the coronoid process, of lesser extent than the olecranon process. Between these bony projections is a curved articular surface, called the sigmoid cavity, for articulation with the trochlear surface at the lower end of the humerus. The sigmoid cavity is a concave curve of about a half circle in extent. This, articulating with the cylindrical surface on the humerus, insures the true hinge joint at the elbow. At the outer side of the ulna, just below the outer lower margin of the greater sigmoid fossa, is a smaller laterally concave surface, the lesser sigmoid fossa, for the articulation of the head of the radius. The olecranon process affords attachment to the capsular and the orbicular ligaments, insertion to the Triceps and Anconeus and origin to Flexor profundus digitorum and Flexor sublimis digitorum muscles.

The shaft of the ulna is prismatic, for most of its length, but is thinner and rounded in its lower third. It affords insertion, near its upper end, to Brachialis anticus and to Anconeus muscles, and origin to Pronator radii teres, Flexor longus pollicis, Flexor profundus digitorum, Pronator quadratus (near its lower end), Extenso ossis metacarpi pollicis, Extensor secundi internodii pollicis and to Extensor indicis muscles.

The lower end of the ulna has a head and a styloid process. The head is rounded below and articulates with the articular fibro-cartilage of the wrist joint. The styloid process is a small projection of bone, extending downward at the extreme lower end of the bone on the inner side. It gives protection to the wrist joint and affords attachment to the internal lateral ligament of the wrist. Near the lower end of the ulna, there are attachments of the anterior radio-ulnar ligament, the anterior radio-carpal ligament, the posterior radio-ulnar ligament and the internal lateral ligament of the wrist joints.

The ulna moves freely on the humerus; the radius moves laterally on the ulna. The two forearm bones are parallel to each other in the anatomical position.

RADIUS

This is a long bone. It has an upper extremity, a shaft and a lower extremity. It is on the outer side of the forearm. The head, at the upper end, is completely circular and is disc-like, for pivoting about itself at its articulation with the lesser sigmoid cavity of the ulna. It has a slight depression on top for articulation with the capitellum of the humerus. Below the disc-like head is a circular narrowing, called the neck. Around the head, but not attached to it, passes a retaining orbicular ligament. The entire head, including its depressed top, is covered with articular cartilage.

The shaft of the radius is thin and somewhat rounded, gradually increasing in bulk toward the lower end. About an inch below the head is a bulge of bone, facing inward toward the ulna, for the insertion of the Biceps muscle. The upper half of the shaft gives insertion to Supinator brevis muscle. At about the middle of the outer side, is found the insertion of the Pronator radii teres and at the lower broadened front is the insertion of Pronator quadratus muscles. The shaft gives extensive origin to Flexor profundus digitorum, to Extensor ossis metacarpi pollicis and to Extensor primi internodii pollicis and (near its lower front) insertion to Pronator quadratus muscles.

The lower extremity of the radius is larger than the head, and is quadrilateral in shape. Its under surface presents a rather large articular face, covered with fibro-cartilage. A slight ridge divides the articular surface into two facets, an outer for articulation with the semilunar bone of the carpus and an inner for articulation with the scaphoid bone of the wrist. There is a styloid process for attachment of the external lateral ligament. The anterior, posterior and outer rims of the articular surfaces give attachment to wrist joint ligaments.

THE ELBOW JOINT

The elbow joint connects the ulna and, to a lesser extent, the radius with the humerus. It is a ginglymus or hinge joint, being the best example of this type of joint in the human frame. This joint depends for its security upon the conformation of the bones involved, especially of the humerus and the ulna. In this it differs from the knee joint, in which ligaments and muscles are the important factors in maintaining the hinge function. The concentrically curved cavity of the ulna's greater sigmoid fossa moves, with exact fit, around the cylindrical trochlear surface at the lower end of the humerus. This arrangement of the humero-ulnar portion of the joint insures the hinge action. In addition, the cup-like depression on the head of the radius moves upon the rounded capitellum of the humerus and allows, but does not insure, the hinged action. To attain pronation and supination of the forearm, (see below), some of the security between radius and humerus is sacrificed.

The Ligaments of the Elbow

They are anterior, posterior, internal and external, which form an encircling capsule, somewhat reinforced at the positions which the named parts occupy about the joint. This capsular ligament extends from the humerus above to the ulna below, except that the external ligament extends to the orbicular ligament of the superior radio-ulnar joint.

The Movements and Muscles of the Elbow Joint

Flexion is *caused* by Biceps muscle, acting from scapula on radius, and by Brachialis anticus, acting from humerus on ulna; by Brachio-radialis and Pronator radii teres, acting from humerus on radius; by Flexor carpi radialis and Palmaris longus, acting from humerus on the wrist; by Flexor carpi ulnaris and Flexor carpi radialis longior, acting from the humerus on the metacarpals; and by Flexor sublimis digitorum, acting from humerus on the fingers. Flexion is definitely *limited*, at about one hundred fifty degrees, by the approximation of the tip of the coronoid process of ulna with the base of the coronoid fossa of the humerus. *Extension* at the elbow is *caused* by Triceps muscle, acting from scapula and humerus on ulna; by Anconeus, acting from humerus to ulna; by Extensor carpi ulnaris, acting from humerus on the metacarpals, and by Extensor communis digitorum and Extensor minimi digiti, acting from humerus on the fingers. Extension is *limited*, at one hundred eighty degrees, by the approximation of the tip of the olecranon process into the base of the olecranon fossa. (It is to be noted that there is no over-extension at the elbow from the anatomical position of the upper extremity, as this position is normally at one hundred eighty degrees; but the needs, in human activities, of extension from a flexed position are so great that the joint is supplied with efficient extensor muscles, the chief of which is the Triceps.)

THE RADIO-ULNAR JOINTS

Although the mechanism of the superior radio-ulnar joint differs materially from that of the inferior radio-ulnar joint, the two actions, pronation and supination, are synchronous. (Also, before discussing their mechanisms separately, we note that the so-called anatomical position of the forearm and hand is one of almost complete supination of the forearm with the palm of the hand facing forward and the thumb outward; and that the anatomical terms of relative direction, such as inward and outward and external and internal, are based on this position of the forearm and hand. This position is due to the fact that anatomists have studied the human body largely from the cadaver, which, in the dorsal position on the dissecting table, has the forearm and hand in this position. On the other hand, in the gymnastic position known as Fundamental Standing, the upper limb hangs at the side of the body, the forearm is midway between pronation and supination and the palm faces inward toward the thigh with the thumb forward.)

A. SUPERIOR RADIO-ULNAR JOINT.

This joint is formed by the head of the radius which twists itself against the lesser sigmoid cavity of the ulna. It is a lateral pivotal joint. The head of the radius is held in its position by an orbicular ligament, which extends from the front edge of the lesser sigmoid fossa on the ulna, around but not attached to the head of the radius, to another attachment at the rear edge of the lesser sigmoid fossa, on the ulna. This orbicular ligament is thus a loop in which the upper end of the radius rotates. A little below the orbicular ligament is a strong band, extending from the edge of the coronoid process of the ulna, downward and outward, to the radius.

B. INFERIOR RADIO-ULNAR JOINT.

As the superior radio-ulnar joint is found at the upper ends of the two bones, side against side, so the inferior radio-ulnar joint is a lateral juxtaposition of the lower ends of the bones, each to each. In the upper joint, a radial head pivots on its own axis against a stationary convex articular ulnar surface; whereas, in the lower joint, a stationary rounded head at the outer side of the ulna contacts a laterally moving concave articular surface at the innerside of the enlarged lower end of the radius. The latter revolves around the former in the lower joint. Thus, in pronation, the shaft of the radius is crossed in front of the shaft of the ulna; in supination, the bones are parallel.

Ligaments of the Inferior Radio-Ulnar Joint

These are anterior radio-ulnar, posterior radio-ulnar short and strong, and an articular fibro-cartilage, between the bones.

Movements and Muscles of the Radio-Ulnar Joints

The movements are pronation and supination.

a. *Pronation.* From the anatomical position, in which the bones are parallel to each other, a forward inward turn of the lower end of the radius, pronation, is made. This consists of a pivoting on the radius' own axis above, a crossing in front of the ulna and a complete transference of the lower end of the radius to a new position in front of and even across the lower end of the ulna. Pronation is *caused* by Pronator radii teres muscle, acting from humerus and from ulna on radius, and by Pronator quadratus, acting from ulna on radius. Pronation is *limited* by the crossed contact of the radius on the ulna.

b. *Supination* from the anatomical position is not possible—the forearm is already in full supination. But, from a position of pronation or in the half-pronation of the forearm in Fundamental Standing Position, supination may be made. Supination is *caused* by Biceps muscle, acting from the scapula on the radius, especially when the elbow is half flexed; by Supinator brevis, acting from humerus on radius, and by Extensor longus pollicis. Supinator brevis is the principal supinator. The motion is through one hundred eighty degrees, and is *limited* by the tension of the fibrous septum between the radius and ulna. (Note: In a right-handed person, screw-driving is an act of supination, and unscrewing is one of pronation.)

In pronation and supination, the wrist, the hand and the fingers turn with and in the same direction as does the forearm. That is, for instance, in semipronation (the gymnastic position), the palm faces inward and in supination (the anatomical position), the palm faces forward.

As noted above (see p. 98), uncomplicated pronation and supination of the forearm is almost impossible unless the elbow is half flexed, to eliminate shoulder-joint rotations. On the other hand, the wrist, the hand and the fingers are necessarily carried with and in the same directions as the forearm. Indeed, the main purpose of the forearm movements is to give variety to the many useful activities of the hand.

THE HAND

The hand, as a whole, (Fig. 20a, p. 90), is the last segment of the human frame to be discussed. For description, it is divided into: A—carpus or wrist, eight bones; B—hand, five bones; and C—fingers or phalanges, fourteen bones.

A. THE WRIST.

The wrist is interposed between the lower end of the forearm—radius and ulna—and the hand—metacarpal bones—and consists of two rows of irregular bones, four bones to each row. The bones are named, for the top row, from radial to ulnar side, scaphoid, semilunar, cuneiform and pisiform; and for the lower row, trapezium, trapezoid, os magnum and unciform. They present palmar and dorsal surfaces for attachment of ligaments and muscles, and articular surfaces for joining each with its neighbor and with the forearm bones above and with the metacarpal bones below. The palmar surfaces give attachment to two palmar ligaments, to an anterior medio-carpal capsular ligament between the trapezium and the base of metacarpal I. The dorsal surface gives attachment to two dorsal ligaments, to a posterior medio-carpal ligament and to a strong transverse ligament. There are also interosseus ligaments between all contiguous bones. The palmar surface of the carpus gives insertion to Flexor carpi ulnaris muscle, acting from humerus, origin to Flexor brevis pollicis, Abductor pollicis and Opponens pollicis, acting on the thumb and origin to Abductor minimi digiti, Flexor brevis minimi digiti and Opponens minimi digiti, acting on the little finger. The dorsal surface of the carpus affords no muscular attachments.

B. THE HAND.

The hand is composed of five metacarpal bones. Each is a long bone with a base or proximal end, a shaft and a head or distal end. The base of metacarpal I, or thumb, has a saddle-shaped articular surface for special articulation with the trapezium of the wrist. The bases of the other four metacarpal bones have rounded articular surfaces, for articulation with neighboring carpal bone or bones, and each has a lateral facet to join with neighboring metacarpal bones. Around the articular surfaces are attached capsular ligaments, and between the bases are found attachments for palmar and dorsal interosseous ligaments. The shafts are delicate and give attachments to muscles. The shaft of metacarpal I is shorter and thicker than the other four and takes a direction outward as well as downward from its articulation with the carpus (trapezium). The shafts of the remaining metacarpal bones are nearly parallel to each other, although slightly diverging at the lower ends. The shafts of metacarpal I gives insertion to Extensor ossis metacarpi pollicis muscle, acting from the ulna, and to Opponens pollicis, acting from the carpus, and origin to dorsal Interossei, acting each on its neighboring metacarpal shaft and on the head of the index finger. The shafts of the other four metacarpal bones give insertion to Flexor carpi ulnaris, Flexor carpi radialis, Extensor carpi radialis longior and Extensor carpi radialis brevior muscles, acting from the humerus;

and to Opponens minimi digiti muscle, acting from the carpus; and origin to Flexor brevis pollicis muscle, acting on the thumb and to palmar and dorsal Interossei, acting on neighboring metacarpal bones. The heads of all metacarpal bones have rounded articular surfaces for union with depressed surfaces on the neighboring bases of proximal phalanges (fingers). Around these articular surfaces are found attachment for lateral posterior and glenoid ligaments. No muscles are attached to the heads of the metacarpal bones.

THE WRIST JOINTS

Text books on Anatomy divide these joints into: a) those between the forearm and the carpal first row; b) those between the first and second carpal row; c) those between the second row and the metacarpals of the hand and d) those connecting the bones each to each laterally. For the purpose of this book, these complex joinings will be considered as one joint, the wrist joint. The ligaments are sufficiently named in the preceding paragraph. They are appropriately located to hold the bones involved together and to limit motion to very little, as between any two bones and yet (as in the region of the spine) to afford considerable total movement. An advantage of this complexity of mechanism is the combination of freedom of motion and yet a high degree of protection from falls and external violence. Yet this protection is not complete as is evidenced in the occurence of sprain of the wrist, which has been experienced, at one time or another, by most persons.

Movements and Muscles of the Wrist

The movements of the wrist are flexion, extension, lateral motions—radial flexion and ulnar flexion—and circumduction. Rotation proper does not exist at the wrist. (It was noted above (see p. 104), that the lateral motions of the forearm gave the rotations required for the hand.) Flexion is freer than extension at the lower part of the joint; extension is freer at the upper part, and the total of flexion is freer than the total of extension. Ulnar flexion is freer than radial flexion.

a. *Flexion* is *caused* by Flexor carpi radialis, Palmaris longus, Flexor carpi ulnaris muscles, acting from humerus on the metacarpus; by Flexor sublimis digitorum, acting from humerus on the fingers; by Flexor profundus digitorum, acting from ulna on fingers, and by Flexor longus pollicis, acting from ulna and radius on the thumb. Flexion is *limited*, at about ninety degrees, by tension of the posterior ligaments of the wrist.

b. *Extension* is *caused* by Extensor carpi radialis longior, Extensor carpi radialis brevior and Extensor carpi ulnaris muscles, acting from humerus on the metacarpus; by Extensor communis digitorum and Extensor minimi digiti, acting from humerus on fingers; by Extensor ossis metacarpi pollicis, Extensor longus pollicis and Extensor indicis, acting from ulna on fingers, and by Extensor brevis pollicis, acting from radius on the thumb. Extension is *limited* by the approximation of the rear of the carpal bones on the articular rim of the radius.

c. *Radial Flexion* or *Abduction* is *caused* by Extensor carpi radialis longior muscle, acting from humerus on metacarpals; Extensor ossis metacarpi pollicis and Extensor brevis pollicis, acting from ulna on metacarpus. Radial flexion is *limited* by the tension of the carpal ligament on the ulnar side. It is possible to only a few degrees lateral to the anatomical position.

d. *Ulnar Flexion* or *Adduction* is *caused* by Flexor carpi ulnaris and Extensor carpi ulnaris muscles, acting from humerus on carpus and metacarpus. Ulnar flexion is *limited*, at about forty-five degrees, by the carpal ligaments on the radial side.

e. *Circumduction* is a combination of the four preceding movements, in series, and is *caused* by the successive action of their movements and *limited* by the ligaments governing the same.

THE FIRST CARPO-METACARPAL JOINT

The joints between the carpus and metacarpus II, III, IV and V were included in the discussion of the Wrist Joint, as a whole. However, that between the trapezium (carpus) and metacarpal I, deserves special mention. It is a ginglymus or hinge joint. The downward laterally faced articular surface of trapezium is concave or saddle-shaped. The base of metacarpal I has an articular surface, rounded and saddle-shaped, facing upward and inward. The neutral position of the thumb (metacarpal I) is outward, downward and a trifle forward, in the anatomical position. These bones are connected by a lateral and a posterior ligament and there are two small sesamoid bones, placed on either side of the joint. Dislocation of this joint is rare but sometimes difficult to reduce because of the saddle shape.

Movements and Muscles of Metacarpal I

These differ somewhat from those of the other carpo-metacarpal joints. The movements are flexion and extension and slight lateral motion.

a. *Flexion* is *caused* by Flexor longus pollicis muscle, acting from radius; by Opponens pollicis and Flexor brevis pollicis, acting from carpus, and by Adductor pollicis, acting from metacarpal III. This motion is often called adduction of the thumb. With complemental action on the ulnar side of the hand, this adduction of the thumb produces a shallow cup of the palm of the hand and is important in many occupations. Man is the only animal with the power of opposing the thumb to the other four fingers. This allows both strong and delicate grasping of objects —the strong hand grip of fraternity brothers and of the office seeker or the delicate work of the watchmaker. It is the position of the hand-ball player and of the ball catcher; of the pen holder or the wielder of delicate hand instruments. Flexion is *limited* by the posterior ligaments.

b. *Extension*, often called abduction, is *caused* by Extensor longus pollicis muscle, acting from ulnar; by Extensor brevis pollicis, acting from radius, and by Abductor pollicis. It is *limited*, at the lateral plane of the hand, in anatomical position, by the lateral ligaments. (When the thumb is in full extension, with the tendons of Extensor longus pollicis and Extensor brevis pollicis tensed, a definite hollow is present just proximal to the base of metacarpal I. This has been called "the anatomist's snuff box".)

THE FINGERS—FOUR LESSER

Three bones or phalanges for each, twelve in all, comprise the skeletal part of this segment. Each finger has a proximal, a middle and a distal phalanx. (Note: The adoption, by Anatomists of the terms, "first, second and third", and by Surgeons of the names, "third, second and first", for these phalanges, leads to confusion. The terms here used avoids this.) Each phalanx has a base, a shaft and a head. The base of each phalanx presents a cup-shaped articular surface to unite with the head of the bone proximal to it. All contiguous articular surfaces are surrounded by capsular ligaments and are covered with articular cartilage. On the palmar surface of the shafts, are insertions for Lumbricales, Interossei and Flexor minimi digiti muscles; on the mid-palmar shaft is insertion for Flexor sublimis digitorum and on the palmar surface of the distal phalanges is insertion for Flexor profundis digitorum. On the dorsal surface of the phalanges is insertion for Extensor communis digitorum, Extensor minimi digiti and Extensor indicis muscles; on proximal phalanges, and on the dorsal mid phalanges is found insertion for the dorsal Interossei.

JOINTS OF FOUR LESSER FINGERS

All of the finger joints are ginglymous, are supplied with glenoid and lateral ligaments and allow flexion and extension.

a. *Fexion* is *caused* by Flexor sublimis digitorum muscle, acting from the humerus to mid phalanges; Flexor profundis digitorum, acting from ulna on distal phalanges; Flexor minimi digiti, acting from carpus on the proximal phalanx of the little finger; and by Lumbricales and palmar Interossei, acting from the carpus and metacarpus on the proximal phalanges. Flexion is *limited* by the approximation of the finger tips with the palm of the hand.

b. *Extension* is *caused* by Extensor communis digitorum and Extensor minimi digiti muscles, acting from humerus on the proximal phalanges; by Extensor indicis, acting from the ulna on the proximal phalanges and dorsal Interossei, extending from the metacarpals to the mid phalanges. Extension is *limited*, at a little more than one hundred eighty degrees, by the palmar interphalangeal ligaments. (Note: Those persons who can markedly overextend their fingers do so because of undue relaxation of the anterior ligaments.)

THE THUMB

The thumb is much shorter than the other four fingers, is situated at a distinctly divergent angle from the hand and is composed of two phalanges, proximal and distal. Each phalanx has a base, a shaft and a head. The joint surfaces are surrounded by a capsular ligament and covered with articular cartilage. The proximal phalanx gives insertion to Flexor longus pollicis, Opponens pollicis, Abductor pollicis, Flexor brevis pollicis and Adductor pollicis muscles; the distal phalanx gives insertion to Extensor brevis pollicis, Abductor pollicis, Flexor brevis pollicis and Adductor pollicis.

It will be noted that, for ordinary handicrafts, the thumb is relatively more important than any one of the remaining fingers. That is, with the loss, by accident, of any of the remaining fingers, the thumb may still have opposing usefulness with the remaining fingers, with but little total loss; whereas the loss of the thumb removes at least half of the usefulness of the hand. (Indeed, most accident insurance companies award to the thumb, a sum equal to the total of the four remaining fingers.)

JOINTS OF THE THUMB

These are hinge joints and are found between the rounded articular head of metacarpus I and the shallow base of the proximal phalanx; also between the rounded head of the proximal phalanx and the shallow base of the distal phalanx. These bones are held together by capsular ligaments. The motions in these joints are flexion and extension.

a. *Flexion* is *caused* by Flexor longus pollicis muscle, acting from radius on proximal phalanx; by Opponens pollicis, Abductor pollicis and Flexor brevis pollicis, acting from carpus on the proximal phalanx, and by Adductor pollicis, acting from metacarpus on the proximal phalanx. Flexion is *limited* by the dorsal portions of the capsular ligaments and is somewhat greater between the two phalanges than between the proximal and the metacarpal bones.

b. *Extension* is *caused* by Extensor brevis pollicis muscle, acting from radius on the distal phalanx; by Abductor pollicis and Flexor brevis pollicis, acting from the carpus on the distal phalanx, and by Adductor pollicis, acting from metacarpus on the distal phalanx. Extension is *limited* by the front portions of the capsular ligaments—that between metacarpus and proximal phalanx— and the ligaments between the two bones.

Some persons, especially in childhood, can voluntarily give opposite motions to the same finger or fingers in the different joints of the same. This is possible because of the special arrangement of the tendons of the muscles which have been enumerated.

KINESIOLOGY OF THE UPPER EXTREMITY

The upper extremity presents some interesting mechanical problems, both from the viewpoint of Fundamental Standing Position and from that of its own structure. In this up-build of standing posture, from the soles of the feet to the top of the head, has been noted the nice balance of the dual support of the lower extremities and of the axial skeleton about the line of gravity. The effect, on this balance, of variations in the relative positions of the segments have been studied. From the soles of the feet to the top of the head, each segment rested on those below and was necessary for the support of the maintenance of the body axis. The upper extremities, on the other hand, have no part in the general support of the body. They are added weights, laterally balanced, to be sure, but still an increase of the load on the supporting structure. So long as these appendages remain *laterally symetrical*, whether in the anatomical or gymnastic position, at the side of the body, or with arm raisings sideways, the upper extremities do not effect posture, the balance is maintained. If, however, there is assymetrical lateral displacements or forward or backward displacements, of any or both arms, there is imbalance of posture. Thus, if the right arm only is raised sideward, there is a corresponding swaying of legs, trunk and head toward the opposite side, unless this is combatted by definite muscular activities, such as was noted in the discussion of the kinesiology of the segments of Fundamental Standing Position. Also, the raisings of either or both arms forward causes a backward swaying which must be controlled by corresponding actions of steadying muscles.

Even though these adjustments may be unconscious ones, on the part of the individual, they are never-the-less actual. Some of these adjustments of axial relationships, due to variations in the positions of the arms will be noted in the study of corrective gymnastic movements in Part II.

However, the upper extremity presents good examples of *muscles,* or muscle groups, *acting over several joints* which deserve study.

I. *Arm raising sideways upward* (Fig. 21), in the standing position, is a combination of elevation of the clavicle, upward rotation of the scapula and abduction of the humerus. The tips of the fingers describe a semi-circle or one hundred eighty degrees at the side of the body axis. This motion has been tentatively divided into:

a—first thirty degrees away from the thigh, humeral abduction only;

b—next ninety degrees, a combination of abduction at the shoulder and upward rotation of the scapula; and

c—The final sixty degrees, an upward rotation of the scapula and elevation of the clavicle.[2]

In an attempt to verify or disprove this, the following X-ray study[3] was made of four consecutive films of the shoulder region of a gymnastically trained individual. The subject was posed with arm at side of the body, at arm carried to approximately thirty degrees from the side of the body, at one hundred twenty degrees and at one hundred eighty degrees, the arm having been carried sideways upward.

Measurement and study of tracings of these films are not conclusive—indeed, one series scarcely could be—but showed some variation from the original proposition.

In the starting position, with arm hanging at the side of the body, the clavicle was at an upward outward slant of eight degrees from the horizontal, the vertebral border of the scapula was angled at eight degrees from the spinal perpendicular and the

---

[2] The writer is unable to trace the origin of this proposition.

[3] This study was made with the co-operation of Dr. William E. Howes, of Brooklyn, N. Y.

axis of the humerus was parallel to the vertebral border of the scapula.

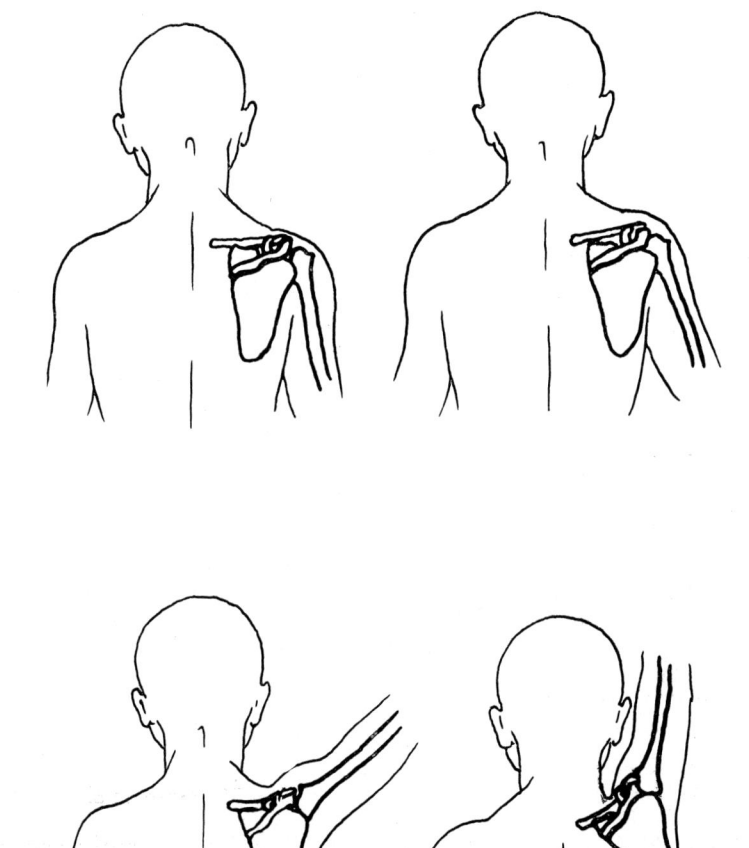

Fig. 21. Shoulder joint and shoulder girdle mechanism, in right arm raising sideways upward (see text).

At thirty degrees of arm raising sideways, the clavicle was raised four degrees from its starting position, the scapula registered four degrees of upward motion and the humerus was at thirty-five degrees from the vertebral border.

At one hundred twenty degrees of arm raising sideways upward, the clavicle had nineteen degrees of elevation, the scapula had added twenty-five degrees of upward rotation and the humerus was in sixty degrees of abduction.

At the vertical arm position (one hundred eighty degrees) the clavicle had lost one degree of elevation, the scapula had lost eight degrees of upward rotation and the humerus had added fifteen degrees of abduction.

Or, stated otherwise, in arm raising sideways upward:

The clavicle was elevated:
    at 30 degree position of arm, 4 degrees
    at 120 " " " " 15 " more = 19 degrees
    at 180 " " " " less 1 = 18 degrees in all

The scapula was upward rotated:
    at 30 degree position of arm, 4 degrees
    at 120 " " " " 25 " more = 29 degrees
    at 180 " " " " less 8 = 21 degrees in all

The humerus was abducted:
    at 30 degree position of arm, 35 degrees
    at 120 " " " " 25 "
    at 180 " " " " 15 " = 75 degrees in all

This single test, made on but one individual, simply confirms the well known composite action of three bones—the clavicle, the scapula and humerus—, and itemizes the movement in these bones. The study will be of value in emphasizing the sterno-clavicular joint as the primal center of action in the movement of arm raising sideways upward.

II. Compound Action of Latissimus dorsi Muscle. This very important muscle (Fig. 22) of the back has origin on the rear crest of the ilium, the spinous processes of the sacral, the lumbar and the six lower thoracic vertebrae, the rear sides of the lower ribs (and sometimes at the inferior angle of the scapula).

FUNDAMENTAL STANDING POSITION 115

From this extensive origin, the fibres of the muscle coverage in an upward outward direction to a tendon, which continues this direction, to an insertion in a groove in front of and near the upper end of the humerus. It therefore affects, by its action, all of the lumbar and half of the thoracic joints, the lower ribs, the clavicle and scapula and the humerus. Acting from its fixed origin, it aids

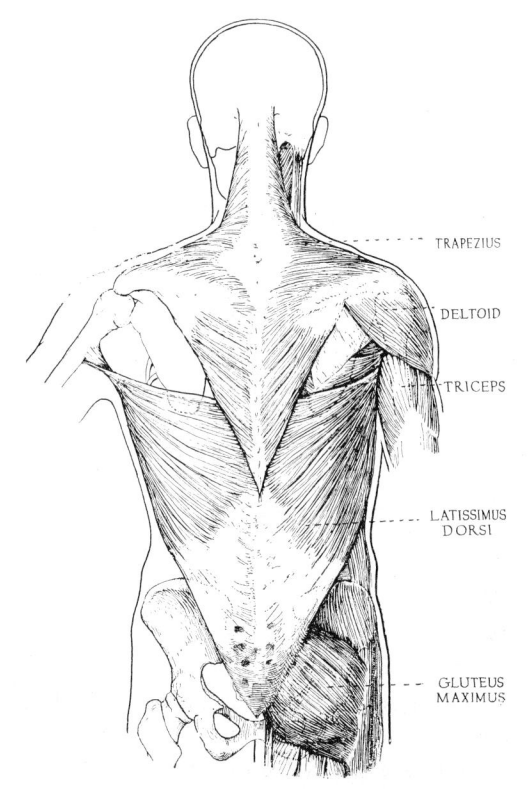

FIG. 22. Superficial muscles of the back, right shoulder and right hip, ⅑ life

the intrinsic respiratory muscles in forcible expiration; it depresses the clavicle, adducts the scapula and rotates the scapula downward and it extends, adducts and inrotates the humerus. This muscle is the principal arm motive power in swimming—whether by the breast stroke or the crawl. In a reverse motion, from the fixed insertion, it laterally flexes half of the thoracic and lumbar spine and lifts the lower trunk, as in climbing.

III. Compound Action of Trapezius (Fig. 22). The Trapezius is a large triangularly shaped muscle, situated at the upper rear of the trunk and the neck. It has origin, from the base of the skull, the ligamentum nuchae, at the back of the neck, and the spines of all the thoracic vertebrae. From this extensive origin, the fibres converge outward and downward, outward and outward and upward to insertions on the clavicle and the acromion process and the spine of the scapula. The action of this muscle may therefore be on the head, the neck, the thoracic spine, the bony thorax and on the shoulder girdle—clavicle and scapula. Or, acting from the origin, the entire muscle adducts the shoulder blade, and the upper part rotates the scapula upward and elevates the shoulder girdle. The two Trapezii play an important part in holding the position of the shoulders in good posture, and either of them is an important factor in maintaining adduction in the scapular action of arm raising sideways upward, while the Serratus magnus (the other upward rotator of the scapula) would tend to abduct the scapula. In reverse action, with the clavicle and scapula insertions as the fixed points, the upper portion of the Trapezius flexes the head and neck laterally and turns the face to the opposite side.

IV. Compound Action of the Pectoralis Major (Fig. 23). This muscle has origin on the inner half of the clavicle, on the sternum and on the cartilages of ribs II, III, IV, V and VI. From this origin, the fibres converge outward downward, outward and outward upward across the front of the axillary space, below the shoulder joint; and the broadened tendon is inserted in the groove on the front of the shaft of the humerus near the latter's top. The muscle acts indirectly on the shoulder girdle and directly on the humerus. Pectoralis major flexes the humerus, adducts the humerus, rotates the humerus inward, abducts the scapula and rotates the scapula upward. In its relation to the shoulder girdle, it is the strongest opponent of the Trapezius. In its action on the humerus, it opposes the extension action of the Latissimus dorsi but aids the adductive and the inrotative power of that muscle. Together with the anterior portion of the Deltoid, the Pectoralis major is one of the principal muscles used in the

FUNDAMENTAL STANDING POSITION 117

forward thrusts of the boxer and of the fencer (in so far as the shoulder mechanism is concerned in these activities). In reverse action, from the humeral insertion, the Pectoralis major is important in pulling up the chest in climbing, just as the Latissimus dorsi pulls up the lower trunk in this activity.

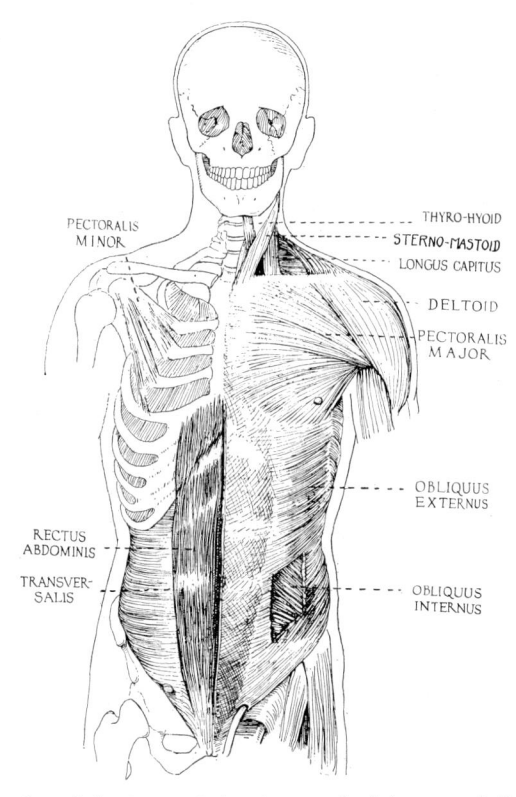

Fig. 23. Muscles of the front of the chest and abdomen, of the left shoulder and left hip, ⅓ life.

V. Compound Action of Biceps (Fig. 20B, p. 90). This muscle, tendinous at both ends and bellied in the middle, acts on the shoulder joint, on the elbow joint and on the radio-ulnar joints. It is situated at the front of the upper arm. It has origin by two "heads"—the long head on the coracoid process of the scapula and the short head on the scapula just above the glenoid cavity. The long head extends outward downward over the head of the humerus (and helps to hold that bone in place) then downward; the short head extends downward; the bellied part constitutes the bulge, which the child likes to show in the elbow flexed position. It extends downward and the lower tendon extends across the front of the elbow joint and is inserted on the radius, at a tubercle which is situated at the inner side and near the top of the shaft of the radius. This muscle aids in flexing (by long head) and in adducting (by the short head) the humerus. It flexes the elbow and it supinates the forearm. The Biceps is one of the principal muscles used in bringing objects toward the body. In reverse action, from a more fixed insertion on the radii, the two muscles are specifically used in the "pull-ups" of the gymnasium.

VI. Compound Action of Triceps (Fig. 22, p. 115). This muscle has three heads, a bellied middle and a single tendinous lower end. It is situated at the back of the upper arm. The heads of origin are the long, fixed at the under-border of the glenoid cavity; the external, attached to the upper external rear of the shaft of the humerus; and the internal, fixed to the lower internal rear of the shaft of the humerus. The upper head is tendinous, the other two have direct muscle-fibre attachments. The bellied part extends down the back of the upper arm. The insertion is by tendon to the olecranon process of the ulna. This muscle, by its long head, adducts the humerus, but its principal action is extension of the elbow. The direct action of the muscle is used in the arm strokes of swimming and in warding off external violence directed to the head and chest. In reverse action, from a more fixed insertion, the muscle is used in hand-standing and in the "dips" on the parallel bars.

(Note: In text books on Physiology, these two muscles, Biceps and Triceps, are frequently used to illustrate the principles of Leverage Action of Muscles. The Biceps action on the forearm and hand is a good example of a lever of the third order—fulcrum at the elbow joint, Biceps power applied on the radius and weight beyond the power or at the center of gravity of the moving arm—the forearm and hand with perhaps some added weight in the hand. The Triceps action on the forearm and hand well illustrates a lever of the first order—the elbow joint between is the fulcrum, the Triceps power is applied behind and the weight or resistance is the forearm and hand, perhaps with the classical dagger grasped in the hand, and is in front.)

VII. Compound Action of Muscles of the Forearm, the Wrist and the Hand (Fig. 20, p. 90). These muscles will not be discussed individually, but will be grouped according to their combined action.

A. Those with *origin* on *internal condyle* of the humerus.

1. Pronator radii teres flexes the elbow and pronates the forearm in writing and using a knife and fork.

2. Flexor carpi radialis and Palmaris longus flex the elbow and flex the wrist.

3. Flexor carpi ulnaris flexes the elbow and flexes and adducts the wrist.

4. Flexor sublimis digitorum flexes elbow, flexes wrist, flexes metacarpus and flexes fingers (middle phalanges of II, III, IV and V).

It will be noted that all of the muscles having origin on the internal condyle of the humerus are *flexors*, some of the elbow and carpus only, some of the elbow, carpus, metacarpus and fingers; and that Pronator radii teres is a flexor of the elbow and a pronator of the forearm.

B. Those having *origin* on the *external condyle* or just above it on the humerus.

1. Supinator brevis extends the elbow and supinates the forearm.

2. Extensor carpi radialis longior extends elbow, extends and abducts the wrist and the metacarpus.

3. Extensor carpi radialis brevior extends the elbow and extends the wrist.

4. Extensor carpi ulnaris extends the elbow, extends and adducts the wrist.

5. Extensor communis digitorum extends the elbow, extends the wrist and extends fingers II, III, IV and V.

6. Extensor minimi digiti extends the elbow, extends the wrist and extends finger V.

It will be noted that all of the muscles having origin on the external condyle of the humerus are *extensors*, some of the elbow, some of the elbow and wrist, some of the elbow, wrist, metacarpus and fingers; and that the Supinator brevis extends the elbow and supinates the forearm.

C. Those with *origin* on the *palmar* surface of the *forearm*.

1. Flexor profundis digitorum flexes the wrist, flexes the fingers (distal phalanges of II, III, IV and V).

2. Flexor longus pollicis flexes the wrist, flexes the carpo-metacarpal joint I, and flexes the thumb.

D. Those with *origin* on the *dorsal* surface of the *forearm*.

1. Extensor ossis metacarpi pollicis abducts the wrist, extends metacarpal I, abducts metacarpal I.

2. Extensor longus pollicis supinates the forearm, extends the wrist, extends and adducts metacarpal I and extends the thumb.

3. Extensor brevis pollicis extends and abducts the wrist, extends and abducts carpo-metacarpal I and extends the thumb.

4. Extensor indicis extends the wrist, extends and adducts finger II.

E. Those having *origin* on *palmar* surface of the *wrist*.

1. Flexor brevis minimi digiti flexes the wrist, flexes finger V (proximal phalanx).

2. Abductor minimi digiti abducts the hand, flexes finger V and abducts finger V.

3. Abductor pollicis abducts the hand, flexes and abducts the thumb (proximal phalanx) and extends the thumb (middle and distal phalanges).

4. Opponens pollicis flexes the hand, flexes the thumb (proximal phalanx).

5. Flexor brevis pollicis flexes the hand, flexes and abducts the thumb (proximal phalanx).

Certain intrinsic muscles of the upper extremity have been omitted in this grouping of muscles with compound action, as they have been included in the discussion of the segmental parts and joints involved.

POSTURE IN STATUARY

It is interesting and instructive to study Posture as portrayed in noteworthy pieces of statuary, available to all in our museums of art. The great masters, for the most part, have been accurate students of the balance of posture and of muscle and other anatomical features of the human frame.

As an example of this study, the *Hercules and Antaeus* of Pollaiuolo, is here chosen (Fig. 24). Antonio Di Jacopo Del Pollaiuolo was an Italian, who was born in Florence in 1429 and died in Rome in 1498. His Hercules and Antaeus is an eighteen inch bronze. The mythical Hercules is "crushing out the life" of the demigod Antaeus. To nullify, in effect, the legend of ever increasing strength, whenever the demigod, Antaeus, touches the ground, Hercules holds up the deity.

The two figures have a common base of support, represented by the feet of Hercules, somewhat separated and at an angle of about forty-five degrees, each to each. The line of gravity passes from the agonized open mouth, downward through the neck, the rear chest, the pelvis and left hip joint of the Antaeus; through the front of the pelvis and between the lower extremities of the Hercules and reaches the base of support about midway between the latter's feet. Body weight seems to be equally distributed about this line of gravity.

KINESIOLOGY OF THE HERCULES

The ankles are dorsi-flexed and the feet are in supination, requiring eccentric contraction of the calf muscles. Both knees are partially flexed, using eccentric contraction of the Quadriceps femoris muscles. The right thigh is definitely flexed, the left but slightly, and the pelvis is but little rotated although slightly tilted downward to the left. The hip joint and pelvic pose are maintained by the tonic contraction of all hip-joint muscles involved. This bronze portrays, in the Hercules, a remarkable example of the extreme lumbar spine extension possible in the human frame. The lumbar spine is in full extension with moderate lateral flexion to the left (the latter necessitated by the hold of the Antaeus somewhat to Hercules' right front). The lumbar spine position is maintained by the strong abdominal and Psoas magnus muscles of Hercules. The thoracic spine is in a considerable degree of flexion and, because Antaeus does not contact

FIG. 24. Hercules and Anteus, by Pollaiuolo, courtesy of the National Museum (Bargello) Florence (for kinesiological analysis, see text).

Hercules in upper front of the torso of Hercules, the latter's chest is not embarrassed. The position of flexion of the thoracic spine is caused by the arm encircling of the antagonist and not by the muscular action of the spine flexor muscles. But the position of the neck is muscularly active. At the cervico-thoracic joint, the neck is in slight extension, and is thus held by the muscles extending from the upper thoracic spine to the back of the cervical vertebrae. The head, on the other hand, is held in marked flexion, probably centered at the occipito-atloid joint; and is due to the downward pressure of the left hand of Antaeus. This would seem to the writer as the only offensive by the Antaeus, exhibited in the statue. Both shoulder girdles are in abduction of scapulae and, because of the life-crushing purpose of Hercules, there should be strong adduction of the muscles of the scapulae (Trapezii and Rhomboids). The arms are in flexion of about forty-five degrees at the shoulder joints, and presumably there is strong concentric contraction of the forearm adductors, especially of Pectoralis major. The elbows are partially flexed, the forearms slightly pronated, the wrists are flexed and the fingers locked behind Antaeus' low back. The muscles, used for these actions, are in powerful concentric contraction.

(Note: It is difficult to see how the hold of Hercules on the low spine could "crush out the life" of his opponent, unless the sculptor believed that the great strength of Hercules could so press his abdomen against the other's abdomen as to embarrass the latter's breathing sufficiently to cause death. The writer suggests that Pollaiuolo's knowledge of anatomy was greater than his physiological information.)

KINESIOLOGY OF THE ANTAEUS

This figure is based on its contacts with Hercules by left thigh against right thigh, by pelvis and low abdomen against right side of pelvis and of abdomen, by right hand against left low chest and by left hand against top of head. From this base, the pelvis is rotated downward; the right thigh is not extended but is carried backward because of rotation of pelvis, using the Glutei and hamstring muscles; the right knee is nearly semi-flexed, using hamstring muscles, the right foot is slightly plantar flexed, using calf muscles. Also, the left thigh is moderately flexed and abducted, the right knee is in full extension and the right ankle is slightly plantar flexed, all of this using but little muscular action, because of simple gravity pull and to a lesser degree because of left thigh contact against right thigh support. But the lumbar spine is markedly extended and the thoracic spine slightly extended, due to Hercules' arm-encircling on the lumbar spine, presumably with abdominal muscle effort to overcome the lumbar extension. The lumbar extension, however, is aided by the push of both arms of Antaeus on his opponent. The flat left hand, is the fixed point on the head of Hercules; the wrist is in extension; the elbow is half-flexed, with offensive action of the Triceps muscle; the upper arm is half-flexed, and the shoulder girdle is abducted, with offensive action of the adductors of the scapula (Trapezius and Rhomboids). The extended right hand is the fixed point on the left chest of Hercules; the right elbow is half-flexed, with offensive action of the Triceps; the right shoulder joint is extended and abducted, with offensive action of Pectoralis major, and the right shoulder girdle is slightly adducted. The upper cervical spine and the head are in marked extension, using the upper Trapezii, Complexii and other extensor muscles. Presuming on the sculptor's belief that the anguish of Antaeus was due to loss of respiratory function, the position of the head, the open mouth and the dilated nares are characteristic features of "air hunger."

The number of statuary masterpieces that could be studied is legion. It is a fascinating by-path.

Chapter 2

OTHER FUNDAMENTAL GYMNASTIC POSITIONS

Introductory. In Chapter 1, the anatomy and kinesiology of the motor mechanism of the human frame was studied. This was presented, body segment by body segment, from the ground upward, primarily to give the picture of the up-build of Fundamental Standing Position. However, anatomical and kinesiological variations from the fundamental positions were discussed, body part by body part, as preparation for the understanding of the mechanism of the corrective exercises given in Part II. In this chapter, eight other fundamental gymnastic positions, valuable as starting positions for corrective exercises, will be studied. As the anatomy, the kinesiology and leverage principles of the human frame have been so thoroughly presented in Chapter 1, these other starting positions may be more briefly described.

A basic principle, in giving corrective exercises, is so to choose a starting position that concentration of effort may be placed upon a body segment with a minimum effort on other human parts. Although Fundamental Standing Position is the starting point of much school gymnastics, of "setting up" drills and of military marching, it is also used in corrective gymnastics. But it does not meet the requirements of localized effort, on the part to be developed and is reserved for later work, after individualized training has prepared the subject for more complicated body activities. The eight other fundamental positions (named below) and the corrective exercises taken from them (described in Part II) meet these requirements and must be understood.

I. FUNDAMENTAL HALF-STANDING POSITION (FIG. 25)

In this position, the weight of the body rests largely upon one foot and the other foot is placed upon a bench with the hip and knee half bent. The base of support is a two-level surface, which is long and narrow. This position, compared with Fundamental Standing Position, necessitates some backward swaying of the entire torso, with extension at the supporting ankle joint and therefore eccentric contraction of the dorsi-flexor muscles, full

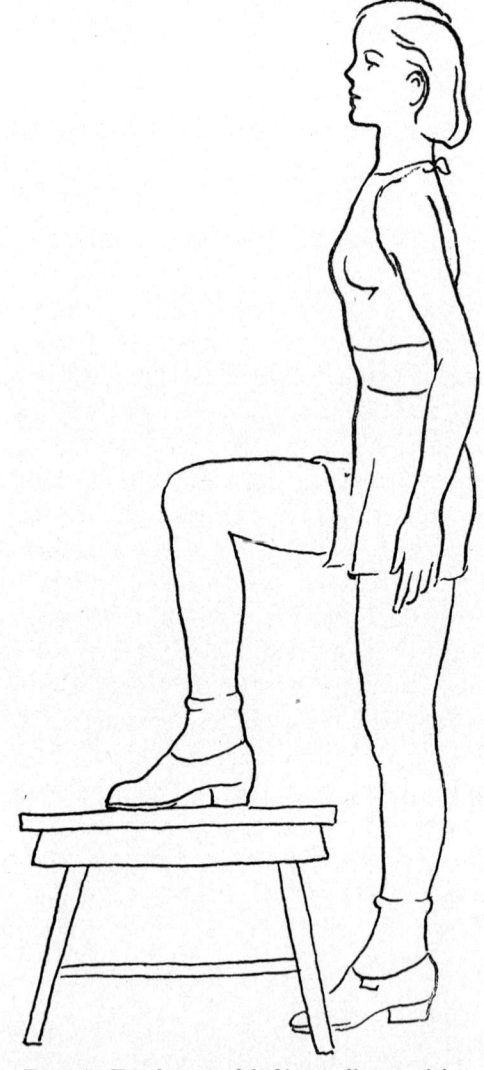

FIG. 25. Fundamental half-standing position

extension of the knee, with added tonicity of all of its controlling muscles and slight flexion of the hip of the floor-supported side. The raised, bench-supported limb requires little or no special muscle activity. The relationships of the pelvis and all parts supported above, do not materially differ from the relationships they have in Fundamental Standing Position. This position, although but poorly meeting the requirements of minimizing effort in body parts not to be exercised, never-the-less is a good one as a starting point for corrective exercises, involving the trunk, the neck, the chest and the upper extremities. It is par-

Fig. 26. Fundamental kneeling position

ticularly useful in the author's "Key-note" derotating exercise for scoliosis (see p. 229).

II. FUNDAMENTAL KNEELING (FIG. 26)

In this position, both knees and legs and plantar-extended feet rest parallel, each to each, on the floor. The base of support is an elongated parallelogram defined by the outer margins of all portions of the body in contact with the floor. As the trunk must

128   BODY POISE

sway backward a trifle, to keep the line of gravity within the base of support, the knees are at just less than a right angle, involving increased tonicity, in eccentric relation, of the extensor muscles of the knee. The hips are very slightly flexed, involving added tonicity of the hip extensor muscles, and the entire framework above is unchanged from Fundamental Standing Position.

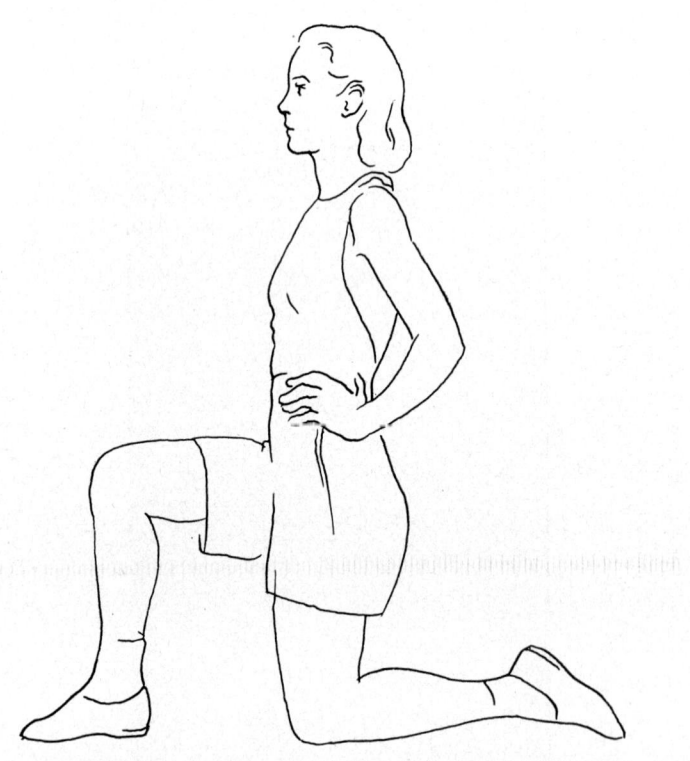

FIG. 27. Fundamental half-kneeling position

From this position, exercises for the correction of faulty posture and for derotation in scoliosis are given (see pp. 180 and 216).

(Note: A variation from kneeling, in which one hip joint is half bent (Fig. 27), its knee joint at right angles and its foot resting on the floor in front, gives a long and narrow base of support, necessitating increased tonicity of all muscles of the down-carried knee and its hip-joint to prevent lateral sway of the body. But this position lessens faulty left lateral deviation of the lumbar curve if the left foot is placed forward and thus allows more concentration on the thoracic curve in derotation exercises for scoliosis—see p. 181.

III. ON HANDS AND KNEES (FIG. 28)

This is a useful starting position in corrective gymnastics. Both palms, knees and feet rest on the floor. The torso is slightly inclined upward (from the rear forward), as the arms are

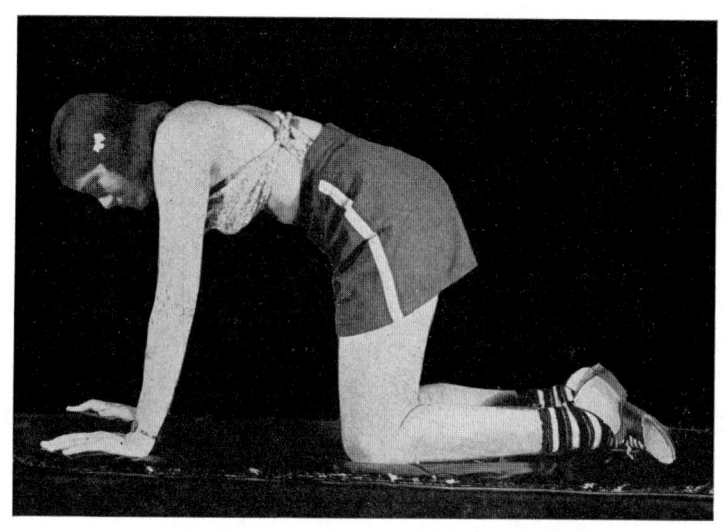

FIG. 28. Fundamental on hands and knees position

longer than the thighs, and the head and neck are slightly more raised from this upward slope of the torso. The base of support is a large stable area, defined by the tips of the fingers, the outer margins of the knees and the toes. Hips and knees are half bent and the feet are plantar flexed. But little added muscular tonicity in the lower extremities is necessary to maintain position in that region. To prevent, sagging of the torso, between its shoulder and hip supports, considerable tonic contraction of the abdominal muscles is necessary. In the shoulder regions, the pectoral muscles are called into play to prevent slump of the upper chest between the shoulder girdle controls. The position of the head and neck is maintained by active contraction of the extensor muscles in this region. (Note: This is the normal position of the head and neck in most quadripeds. It is interesting to note that these animals, especially the horse and allied species, is provided with a massive ligamentum nuchae, at the back of the neck, to ease the effort of neck extensor muscles of these animals.) In corrective exercises, this is an excellent starting position for torso archings and saggings (abdominal muscle development), for leg extensions (glutei development), and arm extensions (back muscle development) and for derotation exercises.

IV. FUNDAMENTAL LYING (FIG. 29)

As its name implies, this is simple lying on the back. It is a relaxed position of the entire body frame and is distinguished from Prone lying. The back of the head, the shoulders and arms, the buttocks, backs of thighs, legs and heels are in contact with the floor. Its base of support is therefore the most extensive available. It is useful as a starting position for straight limb raisings (hip flexor and abdominal exercises), for hip-knee-bend raisings, for hip-knee bend and pelvis lateral rollings (abdominal exercises and lumbar derotation exercises), for trunk raisings (abdominal exercises), and for trunk twistings (derotation exercises).

To prevent strain on the sacro-iliac joints, for most limb raising work, the starting position is modified by a partial hips-knees flexion with the feet resting on the floor just beyond the buttocks.

V. FUNDAMENTAL PRONE LYING (FIG. 30)

In this position, the human frame is prone upon the floor. The chin, chest, arms, abdomen, front of thighs, knees, legs and dorsi of plantar flexed feet are in contact with the floor. The base of support, defined by the points of contact and the lines connecting the extremes of the points of contact, is nearly as extensive as is that of Lying. There is practically complete muscular relaxation.

From this position are taken straight leg raisings (glutei devel-

OTHER FUNDAMENTAL GYMNASTIC POSITIONS 131

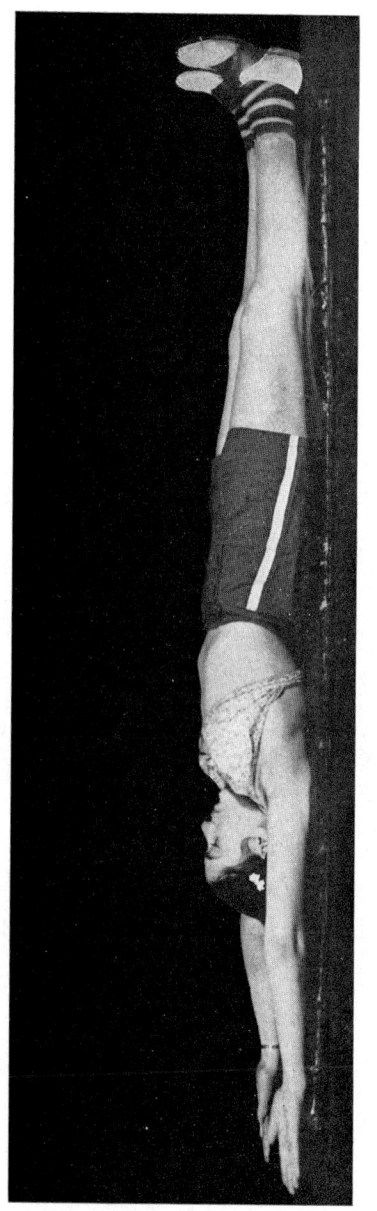

Fig. 29. Fundamental lying position

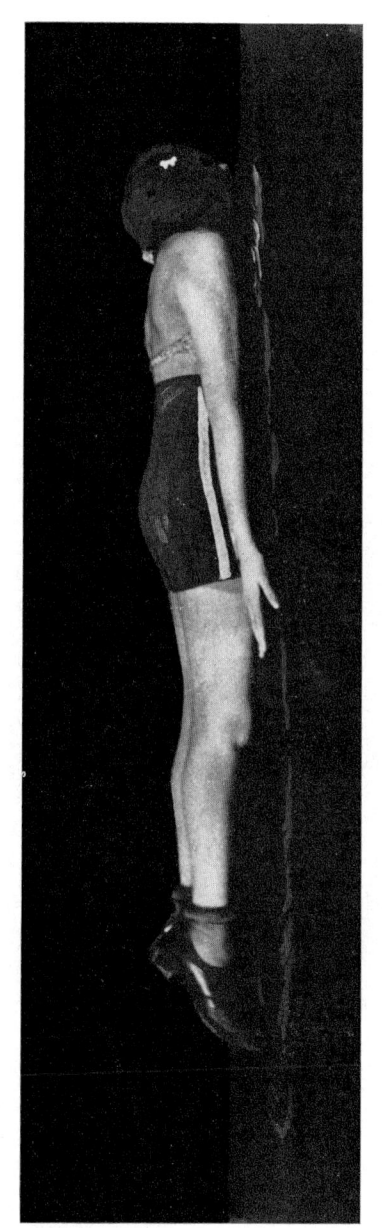

Fig. 30. Fundamental prone-lying position

132   BODY POISE

oping—see p. 41); head and upper trunk raisings (trunk and neck developing muscles—see p. 184) and derotation exercises.

VI. FUNDAMENTAL HALF-PRONE LYING (FIG. 31)

This is a position in which the head (chin) and the entire front of the torso and the arms rest prone on a table, the hips are flexed to approximately a right angle and the feet (toes) touch the

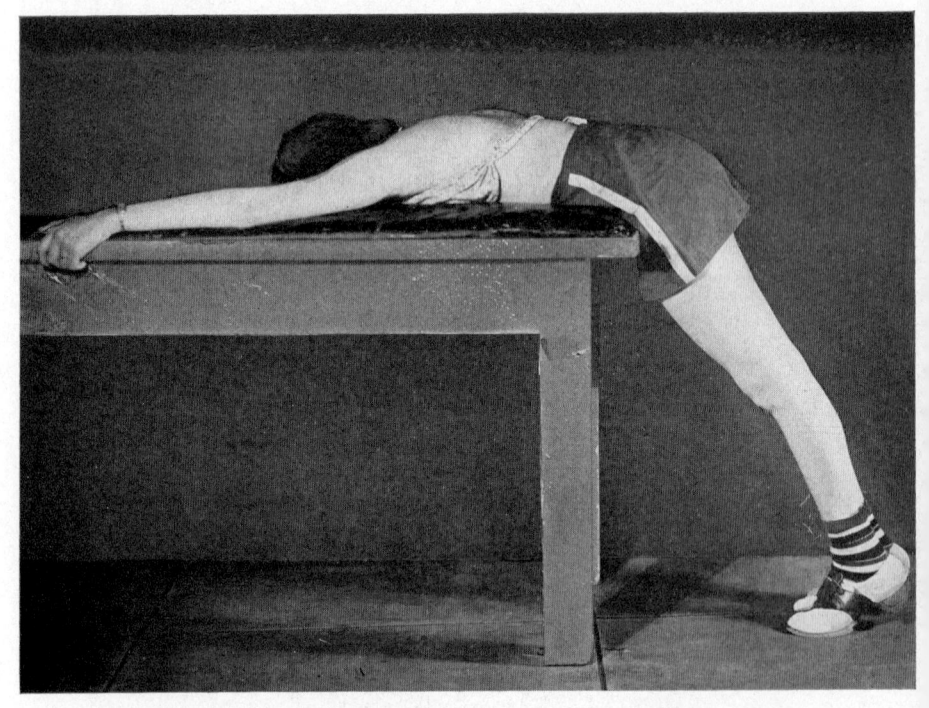

FIG. 31. Fundamental half prone-lying position

floor. (Note: Ideally, the table should be at a height corresponding to the length of the limbs. If, however, a patient's limbs are too long for a perpendicular position, a slight knee-bend is permissable). The base of support is practically that of the chin, arms and torso contacts on the table, although the floor touching of the toes adds some stability. In this position, there is relaxation of the entire body; and because the hips are bent, there is much upward roll of the pelvis and the physiological lumbar curve is almost entirely eliminated. (Note: Because of relaxation and elimination of the lumbar curve, this is a good position for digital examination for painful points in the low back and in the sacro-iliac joint regions).

Half-prone lying at the end of the table is the starting position for straight limb extensions, single and double (Glutei muscles); for parallel limb raisings and pelvis and lower spine twistings

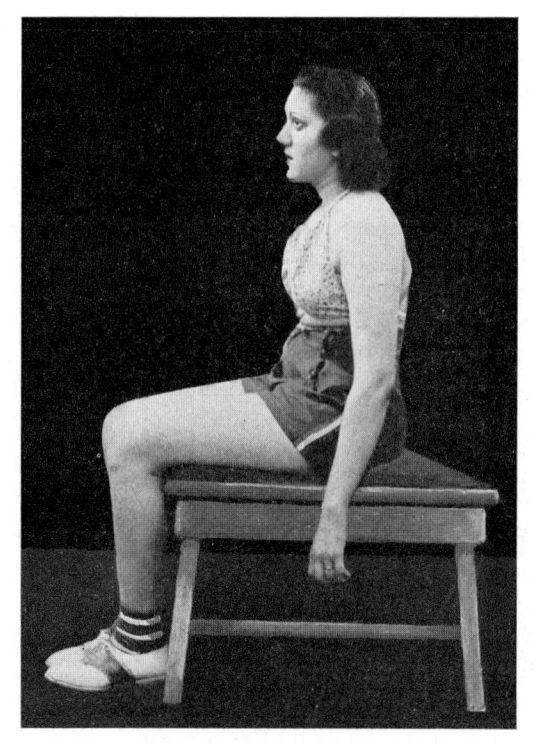

Fig. 32. Fundamental sitting position

(derotation exercise for the lower spine); for upper trunk and head raisings (back extensor muscles), and for upper trunk and head raisings with upper trunk twistings with arms in appropriate assymetrical positions (derotation exercises for thoracic spine).

VII. FUNDAMENTAL SITTING (FIG. 32)

A bench is used, the thighs and buttocks resting on the bench, the feet resting on the floor. The base of support is a two-levelled one and gives stability to all of the body below the pelvis and allowing greater concentration on the torso, head, neck and arms. The position causes the pelvis to be rotated upward slightly with lessening of the physiological lumbar curve.

From this position, are taken the author's foot exercises (see p. 157); alternate and double straight-knee raising (hip flexors and abdominal muscles); alternate and double hip-knee bendings

Fig. 33. Fundamental spring-sitting position

forward (low back muscles); trunk bendings forward with trunk twistings (derotation exercises) and, with feet supported in stall-bars, trunk bendings backward (abdominal muscles).

VIII. SPRING SITTING (FIG. 33)

This position, as a modification of Sitting, was first introduced by the Swedes, more than a century ago, and has been much used since. It is an excellent starting position for the correction of Scoliosis. The thigh of one side is supported on the long axis of a bench, the other thigh-leg-foot is extended far backward, the trunk bends forward and the arms are in various relationships to the trunk—usually, that arm, which is opposite to the backward extended leg, is extended beside the head far forward-upward. This is valuable as a starting position for derotation exercises.

Part II

SOME PATHOLOGICAL DEVIATIONS FROM THE NORMAL INVOLVING POSTURE MECHANISM AND TREATMENT

INTRODUCTION

We are creatures with asymmetrical brains and asymmetrical habits. We write, we answer the telephone, we drive our cars, we sleep favoring one member or one side more than the other. As children, we were perhaps pulled by one arm, we pedalled our "push mobiles" with one leg more than the other, we carried school books on one arm. Later, our athletic games—splendid in developing our muscles, our heart and lungs, our "resistance", our builds—were often assymetrical in their specific effects. A pitcher, a tennis player, a quarter-back, plays and throws intensively with one arm and further develops leg accuracies, abdominal and back controls best adapted to this one-sided delivery. Many gainful occupations, whether sedentary or those of the day laborer or factory worker, are one-sided in their effect on the human frame. In our daily habits of eating, sitting, lounging, leaning or standing, in our clothing itself, especially in shoes, lie influences upon bodily posture, which are no less abusive today for all of the emphasis placed upon them by our schools, periodicals, physicians and life insurance companies. The point seems clear: We are forever engaging in activities which tend toward asymmetry and derangement of our architecture.

So, this book, having indicated the *normal* relationships of the body skeleton and some of the normal mechanical adjustments thereto, in Part I, will, in Part II, elaborate the deviations from the normal as they become pathological. Three basic pathological conditions—Flat Feet, Faulty Posture and Rotary Lateral Curvature of the Spine—will be studied from the viewpoints of (a) pathological deviation from the normal, with methods of measuring and recording; of (b) contributory factors which cause them; of (c) the established general treatment, and of (d) treatment by exercise, including the kinesiology of the exercises given. These three conditions are chosen because (1) they are fundamental divergencies from the normal and because (2) corrective gymnastics play an important part in treatment.

CHAPTER 3

WEAK FEET, FLAT FEET, METATARSALGIA

(Figs. 34 and 35)

Muscle and ligament strains and pains in the long or plantar arches and in the anterior or metatarsal arches of the feet are human complaints as omnipresent as the common cold. And similar may be the patient's reactions to each—irritability, procrastination and neglect.

ANATOMY AND KINESIOLOGY

Although faulty shoes, disease and deforming injury play important parts in causing foot disability, the strains and stresses due to man's *standing position* must remain the *principal factor* in causing weak feet, flat feet and metatarsal derangement. The mechanics of this will be studied.

We have seen (see p. 15) that there are normally three arches to the feet—the long or plantar arch, the metatarsal arch and the dome-like combined arch formed by the two feet side by side. The plantar arch is based at the rear by the tuberosity of the os calcis and at the front by the head of the first metatarsal bone, with the highest point at the scaphoid bone. The metatarsal arch extends from a base at the head of the first metatarsal bone transversely to the head of the fifth metatarsal bone. In a normal foot, the shafts of the three mid metatarsal bones are raised a trifle. (It has been noted—p. 16—that Dudley Morton denies the presence of a metatarsal arch. The writer is prepared to accept this only in so far as the metatarsal heads are concerned, but insists that just back of the heads there is normally a definite raising of the three middle metatarsal shafts, that these shafts may become depressed with actual reverse curve of the metatarsal heads, and that treatment should include support of these shafts—see pp. 149, 151, 152.) The combined mid-foot transverse, formed by the two feet placed side by side, is based by the soft paddings beneath the outer mid border of each foot, rising, from each, inward to an imaginary point midway between the feet. This is the anatomical statement of the three arches of the foot, necessary for description. Too close adherence to them, as

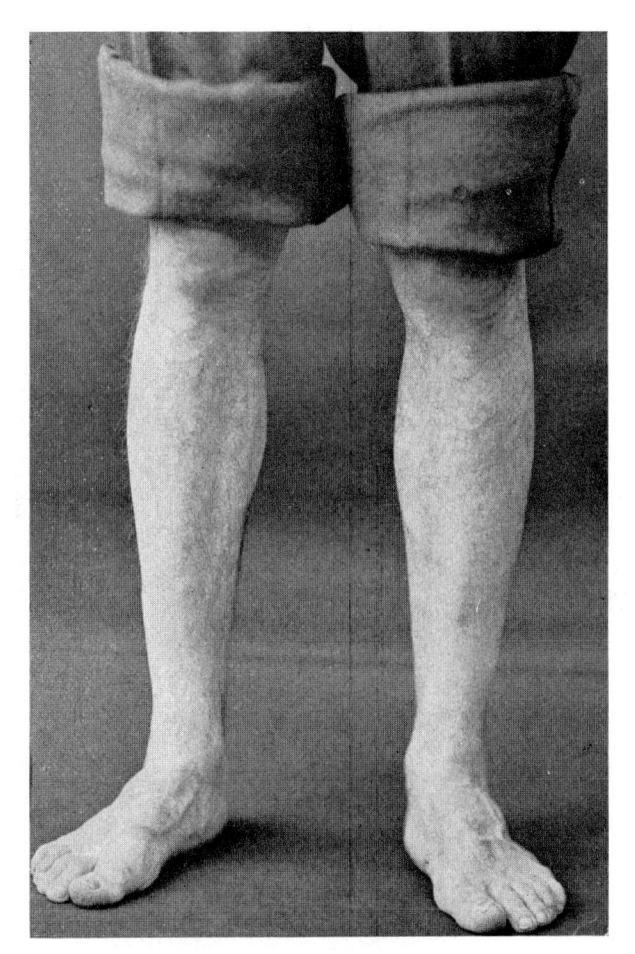

Fig. 34. Moderate weak feet, front view

separate items, has led to errors in treatment, especially in the supportive factor. The sole of the foot is one entity, with a rather complex contour, and should be treated as such (see p. 151).

The arches of the feet, although so easily deranged, are necessary for locomotion, for running, for leaping, for dancing. If the human skeleton were an immobile tower, a permanent flat foot would be a more efficient base. Weight-bearing is the principal active factor in causing plantar weak feet; high heels, abrupt

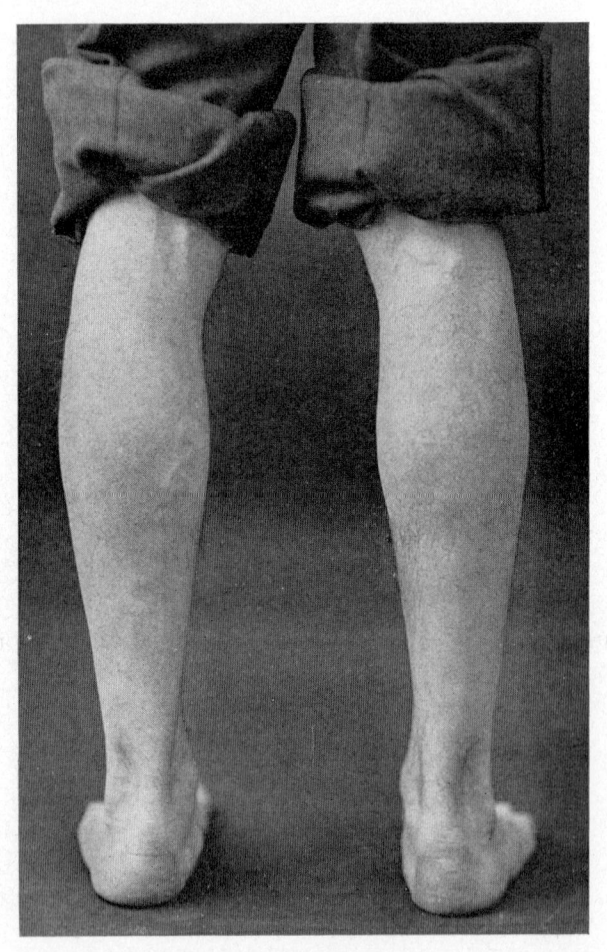

FIG. 35. Moderate weak feet, rear view

shank-sole angle in the shoe and center-pointed toes are the main causative element in metatarsalgia. In weak feet there are, first, strength and loss of the tone of the foot-supporting or supinating muscles—especially Tibialis anticus and Tibialis posticus—with gradual shortening of the opposing or pronating muscles—the Peronei—and stretching and loss of tone of the plantar or sole muscles. Then there is stretching of ligaments—plantar and those between tarsal bones on the inner side of the foot. Finally, we may get true flat foot, a condition in which the articular faces of the tarsal bones become so changed in relation, each to each, that the deformity is more or less set.

In metatarsalgia, there is mis-shaping of the forepart of the foot. The metatarsal heads spread, with stretching of the metatarsal plantar ligaments, and the toes become center pointed. The long and short flexors of the toes and the Transversus pedis muscle become stretched and atonic. Hallux valgus and claw toes may be late complications of this disability. Hallux valgus or center-pointing of the great toe is must commonly caused by the wearing of center-pointed shoes, although the author has shown[1] that an inward divergence of the first metatarsal bone, with structural changes at its proximal end, probably congenital, is present in a small portion of cases. With the center-pointing of the toes is often found a reverse center-pointing of the little toe (Minimi digiti varus). The two deformities give a combined spread of the metatarsal heads, a crowding of the three middle toes and a general mis-shaping of the forefoot, called "Diamond foot". It is obvious that this crowding of the distal ends of the three middle toes creates the necessity for one or more of them to escape from their alignment. They can do this only by finding a place upward.

> ". . . said one little kitten to
> the other two little cats,
> 'If you don't get out of this,
> why, I must.'"

This deformity of the mid toes is called claw toe or hammer toe and may be very painful, certainly is unsightly and becomes difficult in shoe fitting.

EXAMINATION, MEASUREMENT AND RECORDING

Much valuable information may be gained from the sufferer's account of himself—his "history". The duration of symptoms and of disability, the previous treatment, if any, and how effec-

[1] Truslow, W.: Metatarsus Primus Varus. Am. Jl. B. & J. Surg., Vol. 23, 1925.

tive, the shoes worn—all these should be noted. The examiner should inquire into occupation, previous or intercurrent disease and accident. These latter may indeed be contributory factors, and successful treatment will be conditioned on the disposition of these factors.

It is of importance to determine accurately the extent of present functional disability and anatomical deformity. Every patient is an individual problem for therapeutic decision on the relative amounts of protection and of muscle re-education most effective. Each foot requires special study. The examiner notes intrinsic muscle balance: 1) He grasps the waist of the foot with one hand and draws the toes down with the other hand. If this is difficult or causes pain to the examinee, toe extensor muscles are abnormally stronger than toe flexors; 2) the examiner grasps around the ankle, with one hand, the waist of the foot with the other, and turns the forefoot inward. Spasticity and over strength of the peroneii muscles may thus be elicited; 3) with the patient's knee straightened, the examiner dorsi-flexes the foot. Limitation in this movement indicates abnormal shortening of the tendon of Achilles. These muscle imbalances must be overcome before foot efficiency can be attained.

Measurements at the starting of treatment and of progress under treatment should be both easy to make and be constant in their relationship to all of the lines of structure and of force involved in the deformity. The pedigraph, or inked imprint of the soles of the feet, found in some shoe shops, is of value; but the two-dimensional limitations are obvious. Moreover, the writer has found that this projection of the horizontal boundaries of the sole of the foot is far less sensitive to the functional status of the foot than is the Ratio of the Height of the arch to the Length of the foot. From statistics collected upon many consecutive foot problems and of many normal feet, one may state that there is a normal mean ratio between the height of the arch and the length of the foot, in all humans of normal development. This normal runs between the range of 7.7 and 8.3 percent. For example:

A foot measuring 24 centimeters in length and with an arch measuring 1.9 centimeters, has a height-length ratio $(1.9 \div 24)$

WEAK FEET, FLAT FEET, METATARSALGIA 143

or .07916, which is well within the limit stated. If, on the other hand, a foot with the same length (24 centimeters) had an arch height of 1.5 centimeters, the ratio would be $6\frac{1}{4}$ percent, or below normal, and requiring treatment.

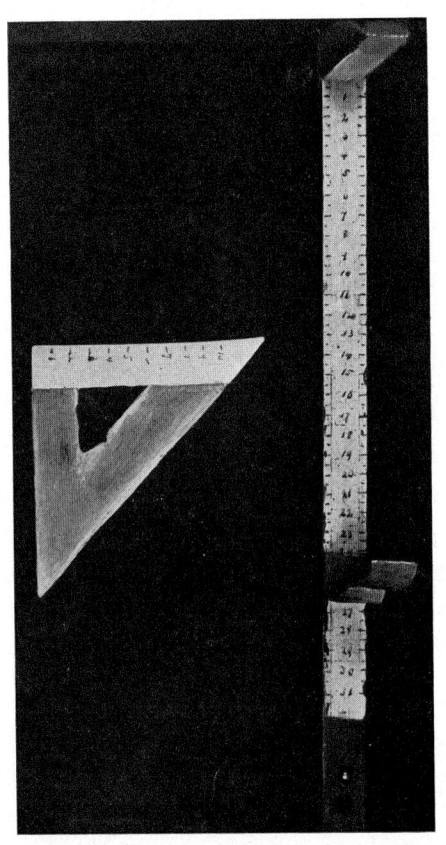

FIG. 36. Simple apparatus for measuring plantar arch ratio. Shoe dealer's sliding rod and draftsman's triangle, both scaled in centimeters.

Simple Apparatus necessary for Recording Height-Length Index

1. Small draftsman's triangle (about 12 x 12 x 18 centimeters), measured off on one of its shorter sides, into centimeters and half-centimeters (Fig. 36).

2. Shoe dealer's foot measuring rod, rescaled from the heel post forward, into centimeters and half-centimeters (Fig. 36).

144   BODY POISE

3. Plain office examining table and bench (the table and bench are for the convenience of the examiner). The examinee sits on the table, the bared feet on the bench.

The measurements taken are:

1. Height of arch standing (Fig. 37): The one to be measured stands on the bench, with feet parallel and apart. The examiner places the measured side on the triangle against the inner side of

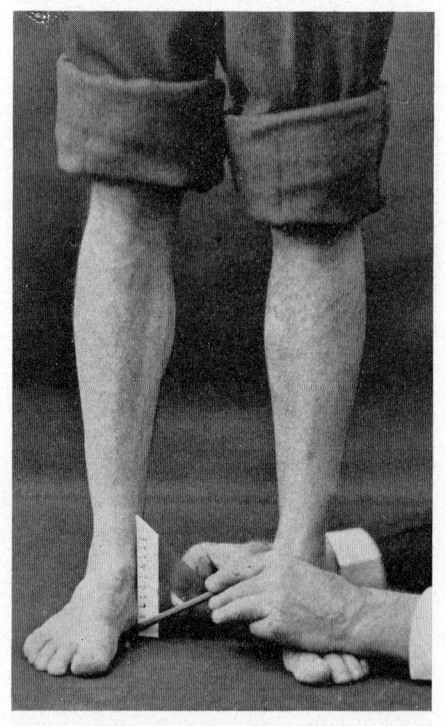

FIG. 37. Measuring height of arch, standing

the foot, at the position of the scaphoid (navicular) bone; then places his pencil point or finger beneath the bone and reads the height on the triangle. He takes and records this for the right foot; then for the left foot.

2. Height of arch sitting: Taken and recorded as in the previous reading, for the right foot; for the left foot.

3. Length of foot sitting (Fig. 38): The measuring rod is placed

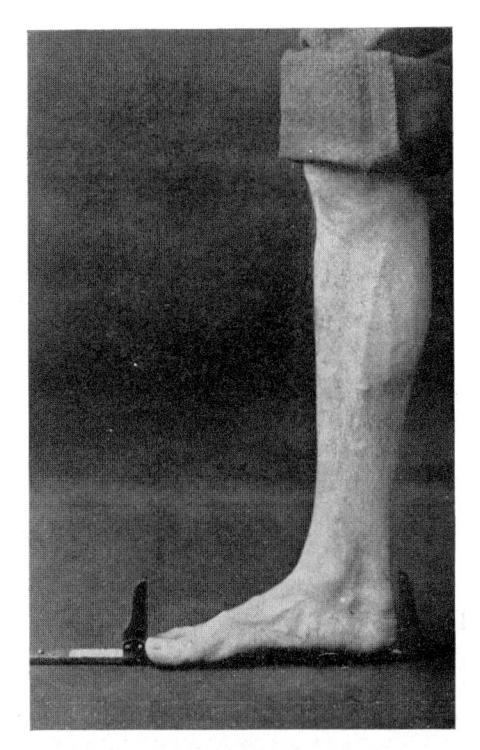

Fig. 38. Measuring length of foot

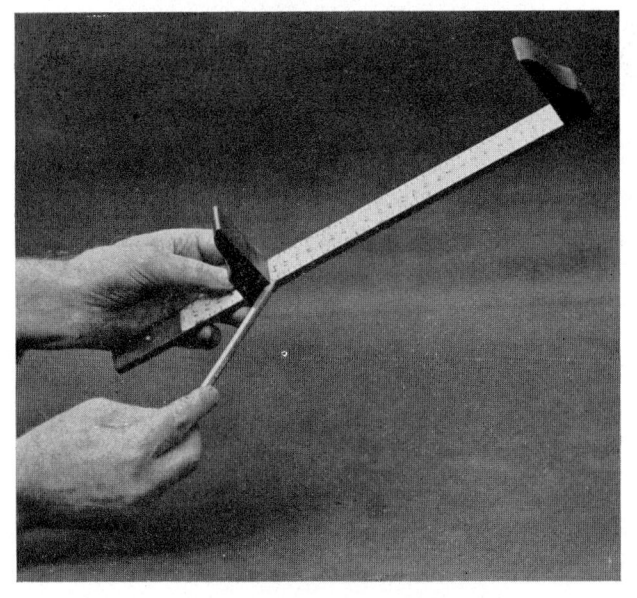

Fig. 39. Noting the foot-length measurement

beneath the foot, the heel of the foot against the heel-post of the measuring rod, and the tip of the great toe over the graduated figures of the measuring rod. The reading at the tip of the great toe is noted and recorded for the right foot; for the left foot (Fig. 39).

Computation of the Height-Length Ratio and Comment

The measured height of the right foot standing is divided by the measured right foot length and the quotient recorded. The same is done for the left foot standing (using the left foot length, if it differs from that of the right foot, which is rare); the same for the right foot sitting and for the left foot sitting.

Thus, a patient's record may read:

R. std. 1.1　L.　.9, equalling .0458　.0375
R. sit. 1.8　L. 1.6,　"　　.0750　.0666
Length 24.　(Each of above figures is divided by 24.)

This may be interpreted as Ratio of right foot standing, $4\frac{1}{2}$ percent; of left foot standing, $3\frac{3}{4}$ percent; of right foot sitting, $7\frac{1}{2}$ percent; of left foot sitting, $6\frac{2}{3}$ percent.

What use does the experienced examiner make of these figures? In the first place they may constitute a record of present conditions, in so far as the plantar arches of the feet are concerned; a comparison of ratios standing with ratios sitting is valuable in the prognosis of the examiner's usefulness (fitness in the military services), and finally repeated measurements, taken later, tell the examiner the results of treatment.

From the writer's experience with these measurements and these ratios, taken on many hundreds of men, women and children, he makes the following *deductions* of their value:[2]

"1. A standing ratio of 8% is a fair average of efficient feet. That is, lacking other harmful factors, a standing range of 7% to 9% may safely be considered as constituting military or civilian efficiency;

"2. The higher the figure of standing ratio (above 8%) the

[2] W. Truslow: How Flat Are the Feet. Medical Times, Sept., 1942.

more is it an expression of existing "hollow foot" (pes cavus), a condition which may become quite as disabling as flat foot;

"3. The lower the figure of standing ratio (below 8%) the greater is the indication of the existence of flat foot (pes planus);

"4. A fairly low standing ratio compared with a high sitting ratio (in the same subject) indicates a temporary muscular weakness—a weakness which the training period of the draftee (for instance) may be expected to overcome. (Note: It is lack of understanding of this principle which has led to the rejection of many registrants who, lacking other disability, would be efficient in military service);

"5. A standing (weight-bearing) ratio, at whatever low level, compared with a sitting (non weight-bearing) ratio which is but little, if any, higher than the standing ratio, points to spacticity of the feet, which may be correctible by prolonged treatment, but which (applied to military registrants) will probably take too long to insure military efficiency. (Note: These cases are well worth caring for in civilian practice);

"6. Marked discrepancy, between the right and the left foot lengths or between the right and the left ratios, calls for more careful study, on the part of the examiner, who may expect to find ante-dating pathological or traumatic causes for such discrepancy."

Thus if, after say two months of treatment, the patient who presented the figures above, at the initial examination, will show:

R. std. 1.5   L. 1.5, equalling .0625 (6¼ percent)
R. sit. 2.0   2.0,   "   .0833 (8⅓ percent),

the examiner will know that the treatment given has equalized the pronation in the two feet and has definitely improved both feet; but that artificial support and corrective exercises must be continued.

The writer does not minimize the importance of other factors, in foot efficiency, elicited by history taking and by careful examination, but insists that the use of this form of foot measurement and of the study of height-length ratios is an important and neglected aid in the evaluation of the effectiveness of military registrants, as well as of individuals in industrial and civilian life.

OUTLINE OF TREATMENT OF WEAK FEET AND THE PLACE OF EXERCISE IN THIS

The care of average foot disability requires:

[3]"1—A careful initial estimate of the needs of the examinee;
2—Means of meeting immediate pain, often by adhesive plaster strapping and sole padding;
3—The temporary use of easily raised office-made insoles;
4—More durable but not bulky instrument maker's insoles;
5—Proper shoes;
6—Faithful follow-through of foot exercises;
7—Gradual withdrawal of the use of insoles."

All of this is important. The crux of the treatment is the proper understanding of the relation, in time and in amount of use, of the principle of support by artificial means on the one hand, and the principle of natural support by intrinsic muscle training on the other (see p. 61). The specialist is consulted because the patient's feet are painful or at least because they easily tire. It would be obviously unfair to withhold from such the immediate relief that is usually possible by employing adequate artificial support. The sufferer might be asked to give ten to fifteen minutes a day to foot-muscle development by exercise and to go about for hours daily with a lack of support. Such a procedure is inadequate. Also, the pain is often severe and the consultor wants relief.

1. The *initial estimate* of the *needs* of the *examinee* is part of the measurement and recording (see p. 143); but should include an estimate of his social, economic and occupational needs and when possible tastes. For instance, the foot-wear needs of the longshore laborer are quite as different from those of the society woman, as are their relative susceptibility to pain.

2. *Pain* brings the patient to the doctor. Temporary *foot strapping* and *sole padding* (Fig. 40) is the best means of meeting this. It should not be continued too long—one, or at most, two weeks of this is usually sufficient. The Examinee sits on the examining table, a foot-stool in front of him, the examiner facing

---

[3] Truslow, W.: Disabled Feet. Med. Times & L. I. Med. Jl., June, 1935.

the examinee, who places the right limb in "side-saddle" position on the table—the thigh, the half-bent knee and the leg supported on the table, while the foot protrudes beyond the table's edge. The pad of one-quarter inch thick piano *felting* is prepared (Fig. 40). Starting from the inner front of the heel, keeping a straight inner edge forward to the rear of the great toe joint, and

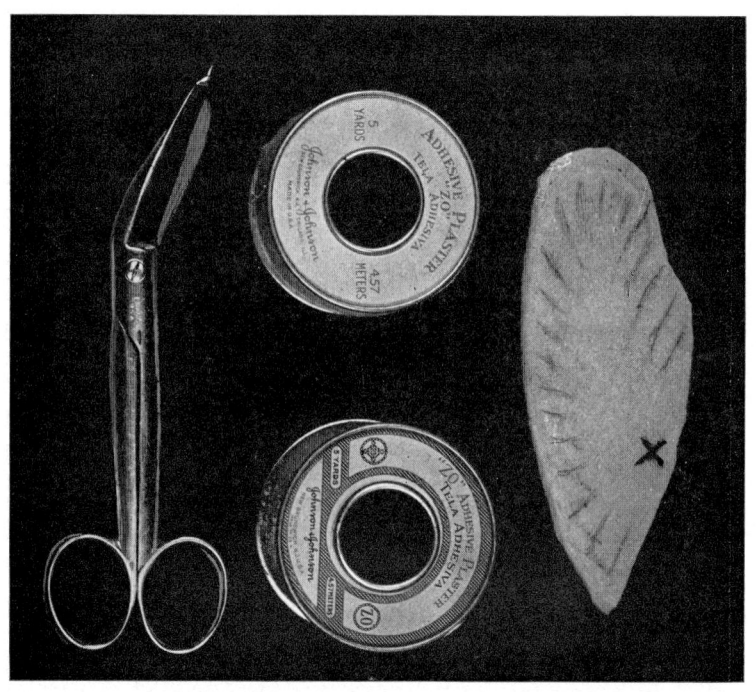

FIG. 40. Materials for foot padding and strapping. One-inch adhesive plaster, felting trimmed and bevelled.

then cutting in further to encircle the great toe joint, then far forward as the bases of toes II, III and IV, the felting shaping then sweeps around backward and inward by a long gradual curve to the starting point. The examinee's position on the table (Fig. 41), with the foot-sole facing the examiner, enables the latter to make this shaping accurately. The top-contouring of the felting padding must now be made. With the shears, the top of the pad is bevelled about its edges in such a way that the highest

150   BODY POISE

point shall be just beneath the navicular bone (Fig. 41), and all edges (except the inner border of the pad, from front of heel to back of great toe joint), shall be sloped gradually away from this highest point. The pad is ready to apply and to be strapped to the sole of the foot. The strapping is made in the form of a slipper (Fig. 41), extending from the great and the small toe

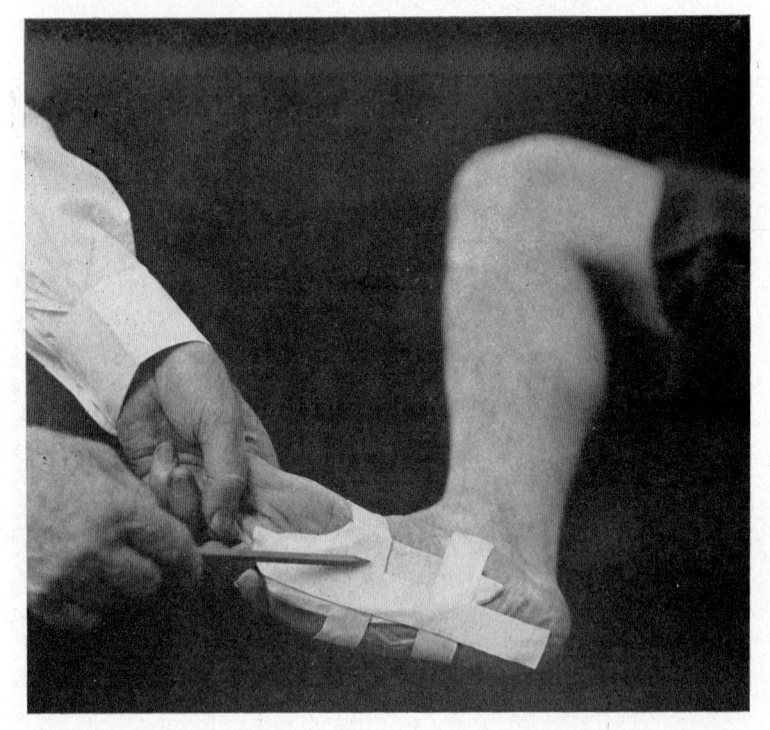

FIG. 41. First step in foot padding and strapping

joints, backward beneath both ankle bones (malleoli) to and including the entire heel; and sideways upward onto the dorsum of the foot (Fig. 42), but leaving an unstrapped lane along the middle of the dorsum, to avoid any embarrassment to circulation in the foot. As the strapping is applied, the patient or an assistant is asked (as in exercise number 4, see p. 160) to hold the forefoot in the corrected adduction position. Rarely is it necessary to carry the strapping on the ankle bones or up the leg. However,

WEAK FEET, FLAT FEET, METATARSALGIA   151

in the presence of marked swelling of the ankles, the strapping should include the ankle bones (as one does for an ankle sprain). Also, with much pain in the calf, the leg is included in the first padding and strapping. A repeated padding and strapping for a second week (and rarely for a third week) is given only if required for pain. The left foot is padded and strapped in the same way,

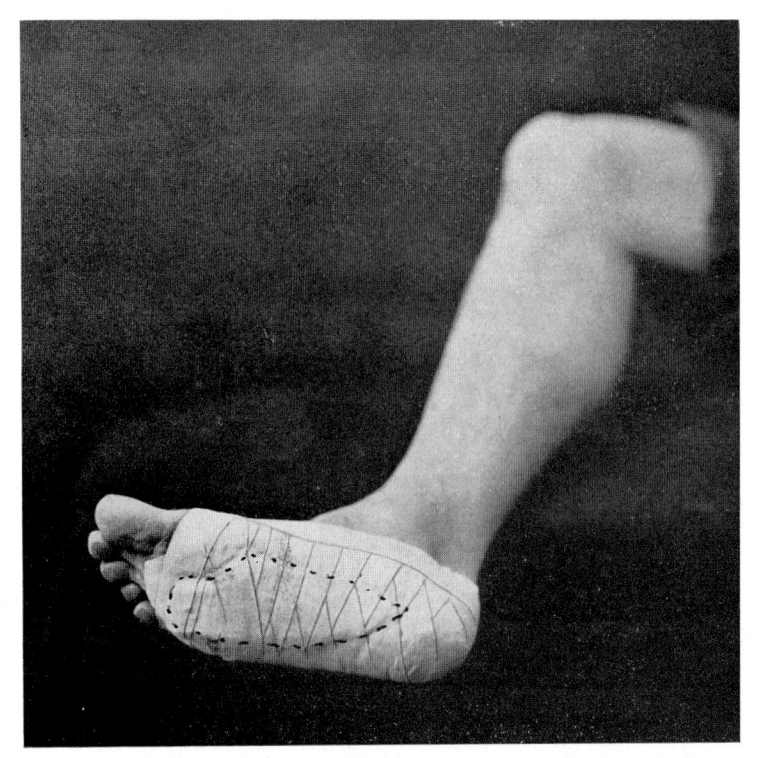

Fig. 42. Sole-contoured "slipper" completed

the examinee shifting his position on the table to the left "side-saddle".

At this point one would condemn the practice of shoe dealers or others of placing a single pad beneath the metatarsal arch, without support to the long (plantar) arch, or of shoe insole padding so placed as to support the plantar arch without caring for the metatarsal arch. Either procedure may give temporary relief, but it does so in each case at the expense of the unsupported arch, which is due to have added strain on that account. We must think in terms of the entire sole. The padding outlined above *evenly supports all arch weaknesses* and adequately *relieves* most *arch pains*.

3. The temporary use of the easily made *office-insoles* is important. The foot-sufferer soon tires of the adhesive plaster strapping although he recognizes the relief from pain which it usually affords, and some skins are intolerant to it. After the padding and strapping, the office-made insole will usually meet the needs adequately. The materials used are thin leather, or two layers of cardboard, and the felting or sponge-rubber or cork contoured filler. The leather or cardboard should be carefully shaped to the insole of the shoe. The felting or sponge-rubber or cork filler should be shaped as in the description of the padding. These parts are put together as follows: The pad is placed beneath the leather or between the cardboards, in such a position thereto as will correspond to the position on the foot described above (see padding and strapping), and is fixed by sewing or using a stationer's riviting apparatus. For women and girls, who now buy shoes with adequate shank reinforcement, no added metal is needed in the insoles unless the patient is heavy. The sewing is omitted or made loosely at the inner edge of the insole to allow added padding from time to time.

4. For men and for children, whose shoes are poorly reinforced at the shank and in women who are heavy, the office-made insole is not strong enough. The instrument maker is asked to make insole *braces* for these (Fig. 43), which should include a minimum metal beneath the padding—just enough metal to "bridge" the shank from the front of the shoe's heel to the front of the downward slope of the shank of the shoe. Carefully contoured sponge-rubber or cork padding is placed above this metal and extends forward to support the metatarsal arch (see padding and strapping) and beneath the top leather insole, which leather insole should exactly fit the insole of the shoe, from the rear of the heel to the base of the toes. But in this, as in the home-made insole, an opening should be left on the inner border of the foot

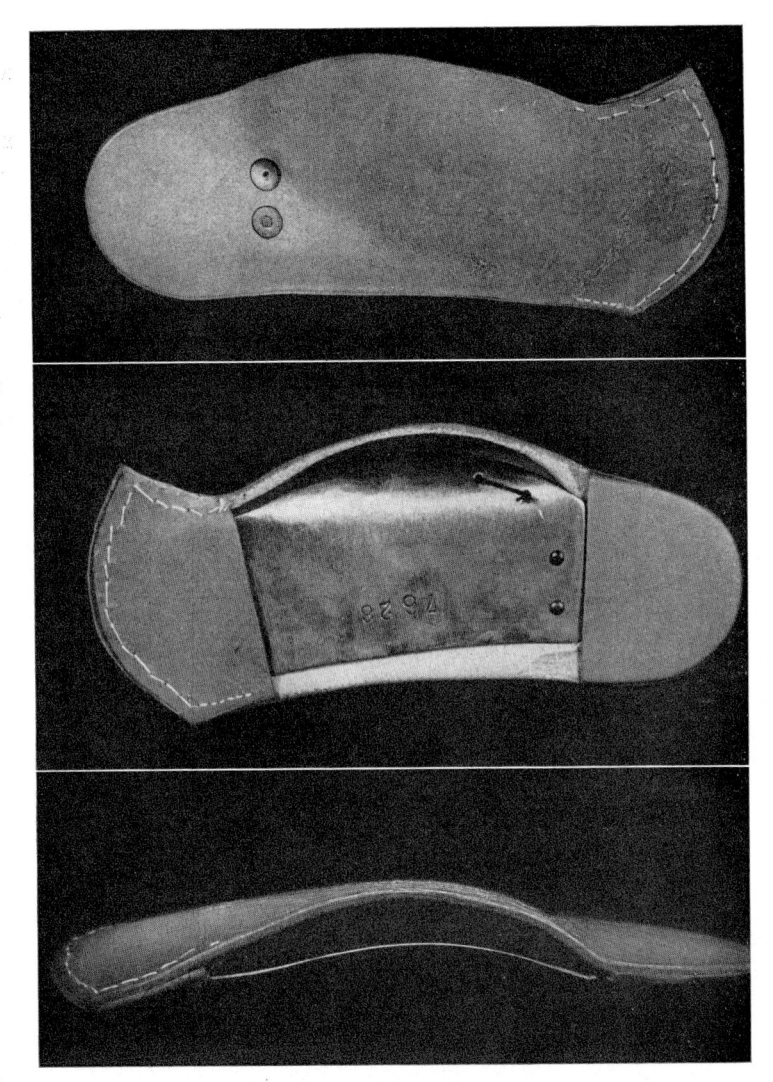

FIG. 43. Insole brace (for right foot). Three views: Minimum metal to "bridge" the shank, contoured sponge-rubber or cork filler, leather insole stabilizer.

brace to allow the doctor to add gradually more and more sponge-rubber or cork filling. The consultant must use his own judgment as when to stop the gradual lifting, but as a general rule it may be stated that when the standing ratio (see measurements) has reached or approximates eight percent, the "mechanical balance" height has been attained.

The all-metal brace, especially the brace with the upward extending flange, to support the heel on the outer side (the Whitman foot brace) is reserved for those cases only in which abduction of the heel is a prominent factor and one which it is difficult otherwise to care for. If it is supplied early in treatment, it is harsh and often very painful. But if the sufferer is prepared for it by initial padding and strapping and by the use for a few weeks of a rapidly raised home-made insole, and if the doctor has been meticulous in so reshaping the plaster cast of the foot (which he sends to the instrument maker) as to insure an accurate fitting to the foot and to the shoe, it is comfortably worn. The writer condemns the practice of sending these foot sufferers to the instrument maker to take his own mold and make his own casting of the foot. Very few are skilled enough to do this in the interest of the patient. Children take well to these all metal heel-flanged braces if their feet have been prepared for them; and in children, as the abducted heel is a very common finding, this form of brace is often indicated.

5. The question of proper *shoes* is of importance. A shoe should not be the factor in causing foot disability and deformity which it often is today, especially those for women and girls. Shoes must be long enough to give full play of the toes and to avoid cramping of the foot at the vamp seam; and they must be roomy enough to do this with the slightly increased bulk which the insoles add to the bulk of the forefoot. The soles of the shoes should be nearly straight at the inner edge. The center-pointed toes must be avoided. An expensive made-to-order shoe, which adequately meets the needs of a given stage in the treatment is a possibility, but it cannot so easily be changed, to meet changes in foot condition, as is entirely feasible with the simple insole carried in a good store-bought shoe. We should expect a shoe to do much toward preventing deformity and disability; rarely is it practicable to get a shoe that will, of itself alone, correct deformity. The heel should be fairly broad. The "Cuban" heel is the

highest that should be allowed for women and girls. Women and girls should be able to—and they can today—get shoes that are good looking and yet not harmful. The shoemakers are increasingly putting out shoes for women and girls which combine style in appearance with avoidance of harmful proportions in make-up. Alas, the latter can and do get many which are distinctly harmful.

6. As soon as relief of pain is attained and proper support and adequate shoes are supplied, the *foot exercises* should be started (see p. 157). They are first used in the sitting position to afford greater accuracy of performance. Later, standing and walking exercises are given. They should be carried through until the feet are able to do their work without artificial aid. To further good tone to the feet, the contrast bath—alternate hot and cold water soaking—and massage are valuable.

7. Artificial *foot bracing* should be *gradually withdrawn* as natural support, by the attainment of muscle efficiency by exercise, is reached. (See diagram in Introduction, p. 61.) As it has often taken months and even years for the feet to become disabled, so the artificial support should not be dropped at once. On the other hand, except in the aged, in those very heavy and sometimes in the lazy, one should avoid indefinitely extended use of the braces. They are needful at first, but we strive for body efficiency; and foot self-sufficiency is attainable in many of the cases—foot comfort in almost all.

Weakness in the feet is not only locally disabling but is often *important as a factor in strains and disabilities above the feet*—in calves, in knees, in hips and in the back.

The crux of the treatment of the more usually encountered foot disabilities is the proper relationship between the principle of artificial support on the one hand and the principle of natural support by intrinsic muscle training on the other. *Mechanical balance must first be attained; then muscle balance must be added to mechanical balance.*

It is interesting to note some similarities between foot strain and eye strain, especially eye strain accompanying lessened efficiency of the muscles of accommodation. Both conditions are gradually progressive. Both are due, in large measure, to increasing functional load placed upon the muscles involved, which muscles become decreasingly able to bear the strain. In both conditions, the age of the patient is a ruling factor, although evidence of both types of strain is found in young adults and even in children. Both are greatly relieved by mechanical support—proper insole bracing in the case of foot strain, accurate eye-glass aid in eye strain. On the other hand, in uncomplicated foot strain, specific muscle training is a definite aid, whereas, the application of muscle re-education is not so well established a procedure in the case of eye strain.

PREVENTIVE MEASURES FOR FEET

1. Keep down excessive weight if possible.
2. Avoid long standing.
3. Excessive weight, long standing occupations, convalescence from illness and maternal confinement call for special support for the feet—firm shanks in the soles of the shoes and sometimes insole foot supports.
4. Working, athletic and playing shoes should be firmly based, ample in size to allow free movement of the toes and with the avoidance of excessively high heels and pointed toes. "Sneakers" should be avoided. Where the rubber-soled shoes are required, the thick crepe sole can be used.
5. Keep shoes in good repair, especially by correcting "runover" heels and thinning soles. Rubber heels are neither corrective nor harmful but they lessen jarring and noise in walking.
6. Use house slippers only temporarily, walking about in them as little as possible.
7. Corns and calluses are almost always caused by undue continuous pressures—the corns on the toes by tight shoe vamps, the sole calluses by pressure of the heads of the metatarsal bones on the skin beneath. Avoid tight shoes and resort to careful chiropody.
8. Avoid out-toeing in walking, as this tends toward flat feet; and stand with the feet parallel.
9. On taking off shoes and stockings, manipulate and gently massage the feet. Bend and twist at the ankle; hand-bend the toes.

WEAK FEET, FLAT FEET, METATARSALGIA 157

EXERCISES FOR FOOT DISABILITY

Starting Position. Sitting with the feet parallel and apart (Fig. 44) (Note: When the feet are correctly placed on the floor,

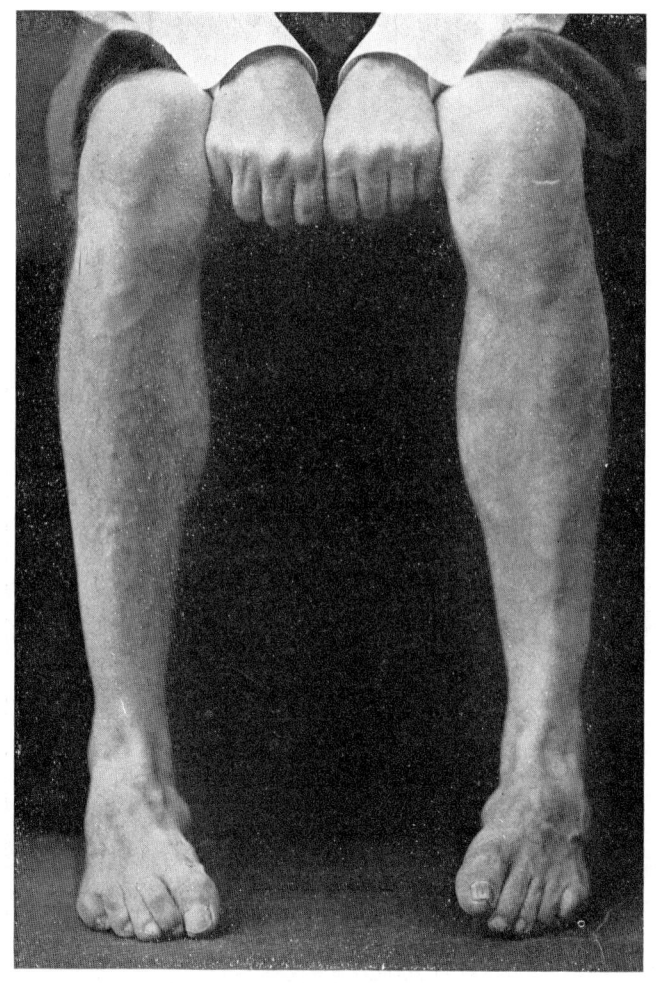

Fig. 44. Foot Ex. 1—Sitting, raise inner borders of feet. (Note fists between the knees to prevent knee movements.)

a squared space exists between the inner borders of the feet. The knees should be directly over the feet. To prevent the knees from moving, in the first two exercises, the patient's fists may be placed between the knees).

Ex. 1. Raise the inner borders of the feet; keep the outer borders on the floor (Fig. 44).

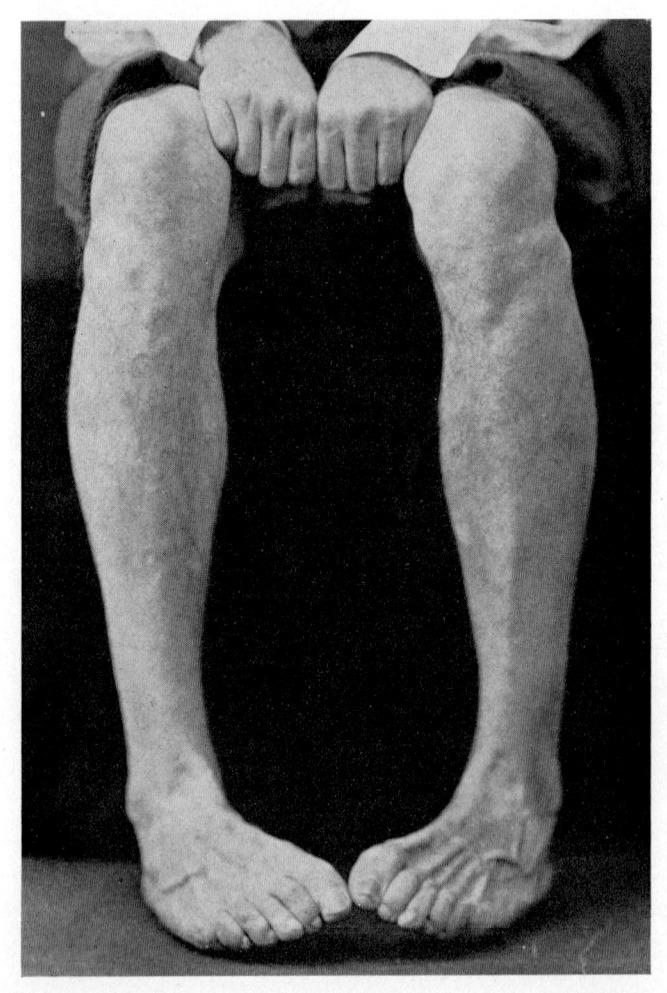

Fig. 45. Foot Ex. 2—Forepart of feet raised, turned inward and inverted with toes flexed. The feet pivot on heels.

Anatomically, this is simple supination of the feet, to develop Tibialis anticus and Tibialis posticus muscles.

Ex. 2. Raise the fore part of the feet, turn the toes under, pivot on the heels until the big toes touch (Fig. 45).

a. Dorsi-flexion of the feet, using Tibialis anticus and Extensor longus hallucis.

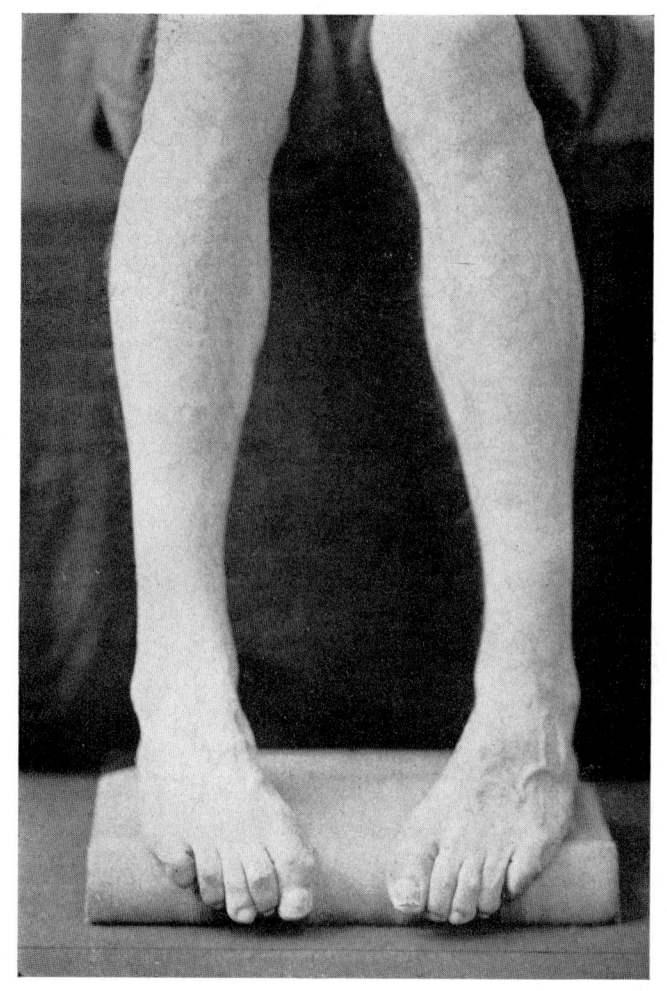

Fig. 46. Foot Ex. 3—The toes are sharply flexed over a block of wood

b. Supination of feet, inversion of forefeet and flexion of toes, using Tibialis anticus and long and short flexors of the toes.

Ex. 3. Place the feet on a book or a block of wood on the floor, toes over the edge (Fig. 46); bend the toes downward over the edge.

FIG. 47. Foot Ex. 4—Right ankle over left knee; forepart of foot is turned upward and inverted, with toes flexed. "Turn the sole heavenward."

Flexion of toes, using long and short flexors of the toes.

Ex. 4. Place the right ankle over the left knee; turn the foot and the sole upward, with the toes under (Fig. 47).

(Note: Do not draw the heel upward toward the calf. The

inverted and toe-flexed foot should be at right angles with the leg.)

Ex. 5. Place the left ankle over the right knee; turn the foot and sole upward, with the toes under.

    a. Abduction and outward rotation of the thigh and semi-flexion of the knee.

    b. Supination of foot, inversion of forefoot and flexion of toes. Same anatomy and kinesiology as in Exercise 2; but using one foot at a time concentrates effort and allows opportunity (in early care of feet) to add guidance to the toe movements, which movements are often erratic until trained.

Later, when Exercise 4 and Exercise 5 are well learned, they may be modified by *adding*

Ex. 6. Right foot flexion and extension (Note: The marked inversion and toe flexions are maintained while making ankle flexions and extensions).

Ex. 7. Left foot flexion and extension.

These exercises add, to Exercise 4 and Exercise 5, flexibility of the ankles and stretch the Tendo Achillis, which are often badly contracted, especially in women, who have developed shortened tendons through the wearing of high heels.

ADVANCED FOOT EXERCISES

Ex. 8. Practice picking up small objects, like jack-stones, with the toes.

This is an advanced refinement of toe co-ordination, to be used by those who show interest in the development of the natural (muscle) supports.

Ex. 9. Practice walking about the room, in bared feet, in "exaggerated club-foot" (Fig. 48).

(Note: This is not to be done until, by the previous exercises, much muscle control has been attained. The writer minimizes weight-bearing exercises, in the early stages of treatment, as putting too much strain on muscles and ligaments and as not conducive to sufficient accuracy of movement. In the "club-foot walking," a soft rug or carpet is the only fair thing to walk on.)

162   BODY POISE

Ex. 10. In ordinary street walking, practice "toes forward"; avoid "toes out."

Out-toeing, in walking, is a definite mechanical cause of flat foot. The out-toed back foot receives a powerfully trans-

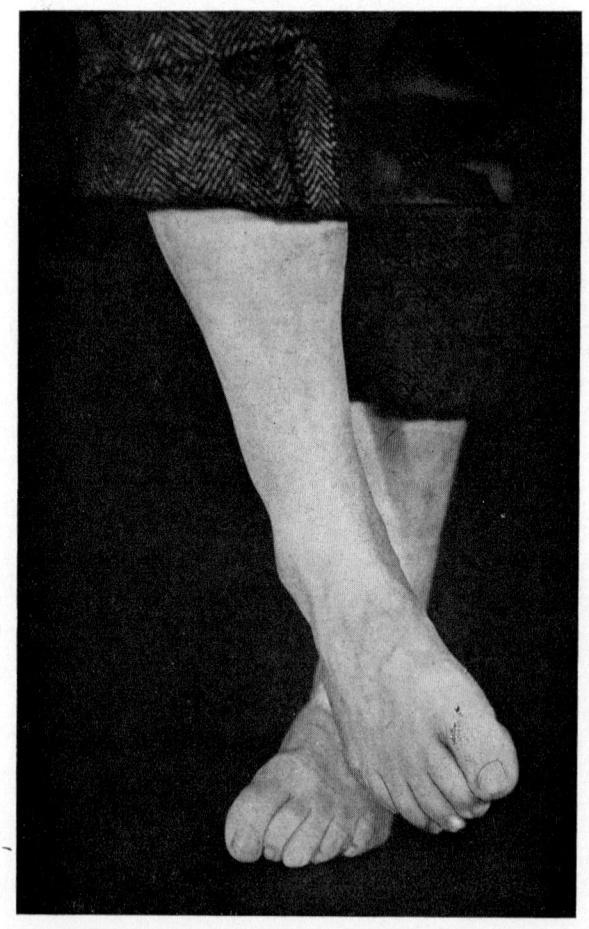

FIG. 48. Foot Ex. 5—"Club foot" walking

mitted leverage action, from the leg above it, as the body is carried forward—and with it the leg—and forces the foot into pronation. In-toeing, in private, and forward foot placing, on the street, obviates this.

SPASTIC FLAT FEET

In describing the mechanism of weak feet (p. 140), it was stated that there is loss of tone of the supinating muscles and shortening of the opposing muscles. In spastic flat feet, this imbalance is accentuated and tends to become permanent if not vigorously treated. The spastic foot presents much greater strength of abductor and evertor (peronei) and of the toe extensor muscles as compared with the adductors and invertors (tibials, etc.) and the toe-flexor muscles. The foot is held in a definite position of abduction, with a flattened sole and often with over-extended toes. The tendon of Achilles may or may not be shortened. The name spastic is given because any attempt at manual reduction of the deformity, that is, of untwisting of the abducted position, is accompanied by spasm and painful resistance of the tightened abductor muscles. Those suffering from this condition are often awakened from sleep by the pain of this spastic contraction of the shortened tendons.

There are varying degrees of this condition, from a simple manually resistant imbalance to an almost fixed deformity in which the tense raised tendons are easily seen on the dorsum of the foot. Although true arthritis may be a complicating factor, it is present in far fewer cases than is generally supposed. Candid but exaggerated muscle imbalance is the principal mechanical condition, and, as such, points the way to treatment.

Whether of mild form or well advanced toward fixation, the *treatment* consists in the restoration of muscle balance. First, the tightened and shortened abductor and toe extensor tendons must be overcome; then the weak and lengthened muscles and tendons must be toned and strengthened. When these are accomplished, there remains the care of the plantar and metatarsal weak feet, as outlined in previous paragraphs.

To overcome the tightened tendons, two procedures are available. If the feet are very resistant, it is advisable to give an anaesthetic, to untwist the feet by forcible manipulation and then to encase them for two to three weeks in a plaster of Paris dressing, extending from just below the knees to the tips of the toes, the feet and toes being held in such corrected position as has been

164   BODY POISE

attained. On removing the plaster dressings, the manipulation (to be described) is started and supportive foot braces applied. The all-metal Whitman foot brace is often necessary for this. Sometimes adhesive plaster strapping is also used.

Vigorous manipulation, perhaps with foot strapping, meets most needs of spastic feet. Surgeons differ somewhat as to the

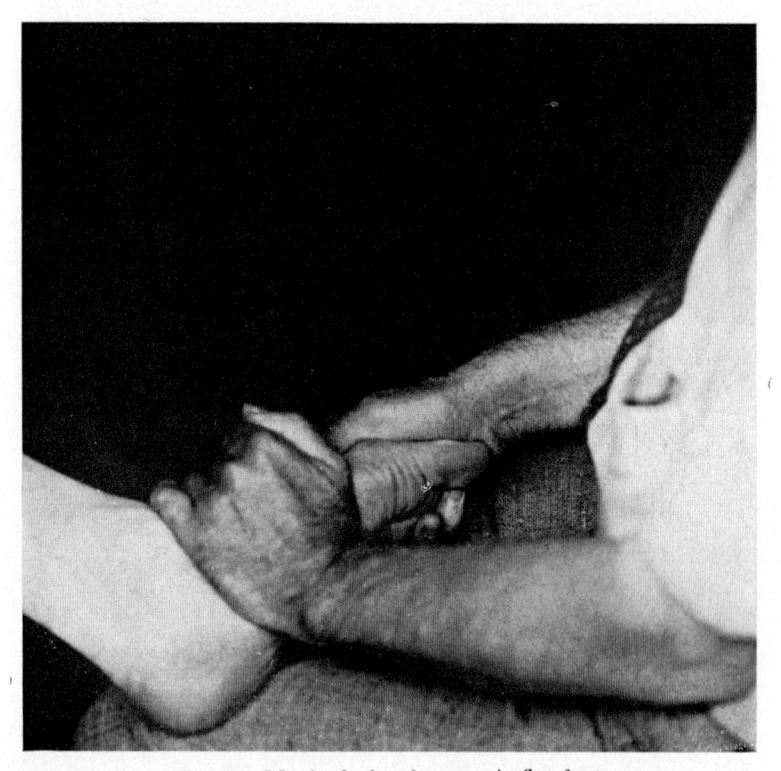

FIG. 49. Manipulation for spastic flat foot

precise procedure. The following has been found effective. The patient sits on a table, the operator on a chair, the position of which is changed for each foot. To untwist the *left foot*, the patient partly bends his left knee and so adjusts his entire limb that the sole of the left foot will face the chair-sitting operator. The outer side of the left heel is placed on the operator's left knee to fix the heel. The operator then places the "heel" of his left hand at the inner mid-foot of the patient, and grasps the dorsum of the forepart and of the toes of the patient's foot, with his right-hand (Fig. 49). Then begins the untwisting: a) the knee holds the heel; b) the left hand holds the mid-foot; and c) the right hand pulls on the forefoot. The forefoot must be turned downward and inward and the toes flexed as much as possible. The operator's active forces begin easily, giving confidence to the patient, who dreads the expected pain, and gradually increases his, the operator's, opposing forces—the hold and the twisting pull—and keeps this up for perhaps five minutes. The *right foot* is manipulated in a similar manner, but the operator changes the position of his chair so as to face the sole of the right foot; and he places the heel of the foot on his right knee, uses his right hand for the mid-foot hold and his left hand for the forefoot and toes twisting pull. He uses his increasing forces slowly but progressively, as when caring for the left foot. Sometimes it is better to give two to three minutes to the left foot, then two to three minutes to the right and to repeat this on both feet.

It is remarkable how much reduction of the foot deformity is accomplished by this slow gradual motion. Of course, this must be repeated. The writer has found it wise to give three such manipulation treatments in the first week, two in the second and one in the third week. If an intelligent and reasonably strong relative or friend can be found to give the treatment daily, some of the work of the experienced operator may be omitted. The patient cannot do this properly to himself.

As was stated above, when abductor tenseness has been overcome, the foot exercises outlined in this chapter, especially number IV and number V, should be used to establish the natural forces. Also proper foot support must be carried on until true functional balance may replace the artificial support. But it cannot be said too strongly that even spastic flat feet may become efficient feet, and that nothing short of vigorous and follow-through treatment can accomplish this.

SUMMARY

1. Weak and strained feet and metatarsal strain cause much pain and disability; and may be responsible for strain and dis-

ability above the feet and are potent factors in faulty body posture.

2. Superimposed weight and local muscle insufficiency constitute the basal causes of foot strain.

3. Ill fitting, tightly cramping and poorly balanced shoes are contributory factors.

4. The care of foot strain and foot disability is protective and muscle re-educative.

5. Initial padding and strapping are the best means of relieving pain.

6. Artificial bracing is valuable at first, but should be gradually lessened and usually discarded when local muscle efficiency becomes adequate.

7. Home-made insoles, without metal, are valuable at first. Instrument maker's braces, with a minimum of metal base, are more durable and are used for heavy persons.

8. Muscle development is necessary to attain foot self-sufficiency. Exercises are first given in sitting positions for better accuracy of performance, but may progress to standing and walking exercises.

9. Measurements and recording are employed to obtain an initial estimate of the condition and to guide continuance of treatment.

10. Spastic flat foot is treated by special manipulation and simple weak foot follow-through.

Chapter 4

FAULTY POSTURE—FUNCTIONAL ANTERO-POSTERIOR AND FUNCTIONAL LATERAL DEVIATIONS OF THE SPINE

DEFINITION AND MECHANISM

In this chapter will be discussed certain deviations from the normal poise of the spine and trunk, which become habitual and in which muscle weakness is a leading factor. In this group no lesion or disease, other than muscle weakness is demonstrable, although certain contributory mechanical defects, to be discussed below, may be factors. For clearness of presentation, normal posture is defined; then variations in the antero-posterior plane of the body—the "kangaroo" and the "gorilla" types—, and finally functional lateral curvature of the spine will be analyzed. Functional lateral curvature is included in this chapter because its cause is similar to that of antero-posterior deviations from the normal and because the treatment of these two functional conditions is practically the same. The mechanism and treatment of structural scoliosis are so radically different, however, as to warrant a separate chapter.

1. Distinguishing features[1] of the *normal type* (Fig. 50) are: a) the line of gravity of the body passes through important pivotal points; b) the pelvis is balanced in equilibrium on the heads of the femora, its position being indicated by the angle which the plane of its inlet makes with the horizon, described as fifty to sixty degrees, and there is no lateral tilt; c) the head is level, the curve of the thoracic spine is gently rounded backward, the curve of the lumbar spine is gently rounded forward, there is no lateral deviation of the spine, the sacro-iliac synchondroses have no posture strain, and all abdominal and pelvic viscera are in their normal relationships; d) the femora are in neutral position between flexion

[1] Dickinson, R. L., and Truslow, W.: Averages in Attitude and Trunk Development in Women and their Relationship to Pain. Jl. Am. Med. Assn., Dec. 14, 1912.

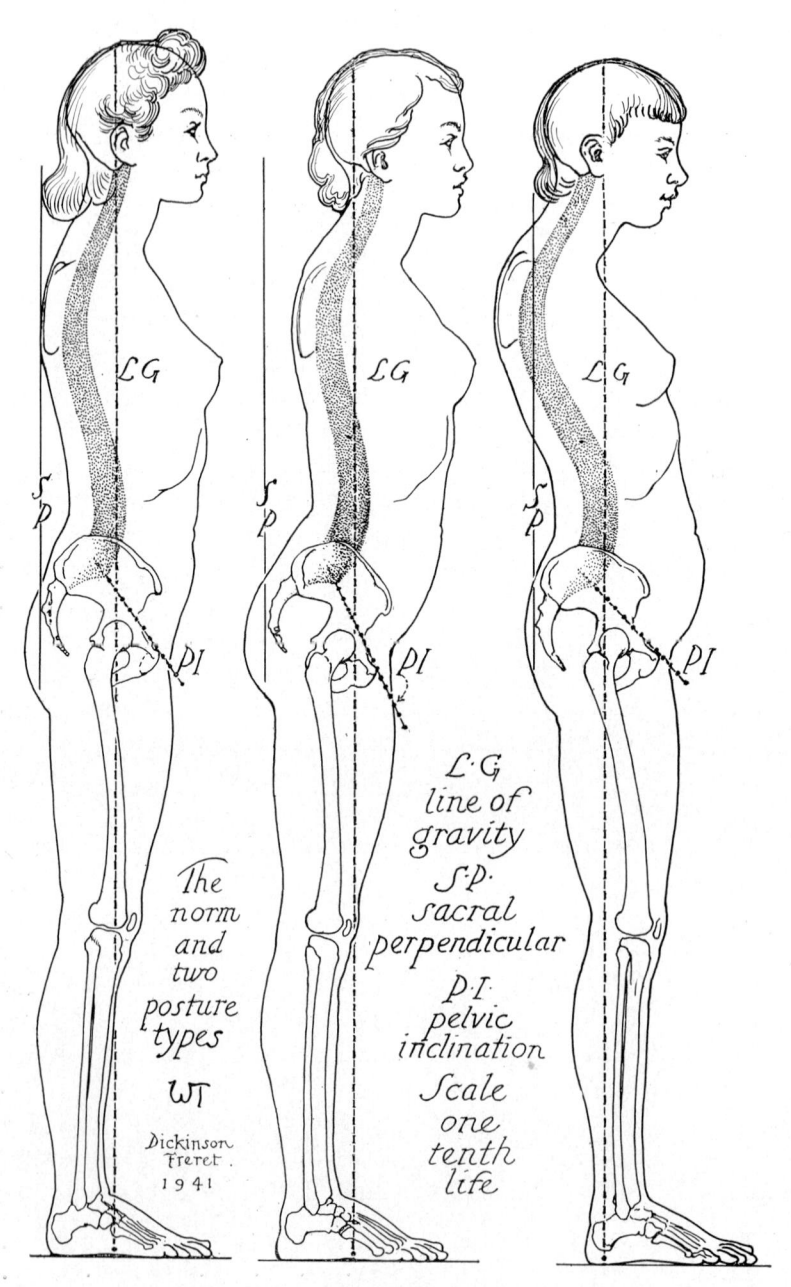

FIG. 50. Three types of posture, side view, $\frac{1}{10}$ life

and extension, the line of gravity (viewed from the side) passes just back of the knees, insuring proper balance of the legs between flexion and extension, and in front of the ankle joints, insuring equal distribution of body weight between the heels and the balls of the foot; e) because of antero-posterior balance at important pivotal points, in standing, there is no muscular strain; f) the rear sacral perpendicular line touches two points, the rear of the thoracic curve and the groove between the buttocks.

2. Distinguishing features of the *"kangaroo" type* (Fig. 50) are: a) the line of gravity of the body passes back of most of the pivotal structures of the trunk and in front of most of the pivotal structures of the lower extremities; b) the pelvis rolls (or rotates) forward downward, the plane of its inlet making with the horizon an angle more obtuse than that of the normal type; c) the head is apt to tilt upward, the mid spine is somewhat flattened and carried in front of the sacral perpendicular, the curves of the low thoracic and of the lumbar spine slant downward backward, the sacrum has a tendency to rotate forward at the sacro-iliac synchondroses, straining these joint mechanisms, the Erector spinae group of muscles are stretched and strained, the distance between the sternum and pubis is increased and there is a tendency to displacement of the abdominal and pelvic organs forward downward; d) the femora tend to a flexed relationship to the pelvis, causing a stretching and strain to the buttock muscles, a shortening of the hip-flexor muscles and a relaxation to the front thigh muscles, the line of gravity passes in front of the hip joints and of the knee joints, the latter giving knee-back deformity and stretch and strain upon the hamstring muscles, and superimposed body-weight tends toward the heels, causing the individual to walk heavily upon the heels; e) because of imbalance at pivotal points, there is muscle strain and stretching of Erector spinae muscles, of the buttock muscles, of the hamstring muscles and of the anterior tibial (plantar arch supporting) muscles and a relaxation of the low abdominal muscles, of the front thigh muscles and of the Peroneal muscles of the leg; f) the rear of the chest is carried in front of the sacral perpendicular line.

3. Distinguishing features of the *"gorilla" type* (Fig. 50) are:

a) the line of gravity of the body passes in front of most of the pivotal structures of the trunk and back of most of the pivotal structures of the lower extremity; b) the plane of its inlet making with the horizon an angle more acute than that of the normal type; c) the head tilts downward, both curves of the spine are exaggerated, the thoracic carried backward and the lumbar forward, the backward rotation of the sacrum strains the sacro-iliac joint mechanism, the distance between the sternum and the pubis is decreased, there is stretching of the Ilio-psoas muscles, relaxation of the abdominal muscles and backward displacement of the upper abdominal viscera, with forward displacement of the lower abdominal viscera and backward displacement of the pelvic viscera in women; d) the femora are extended in relation to the pelvis, but owing to Ilio-psoas strain and tension of the Y-ligaments at the hips, the thighs are carried forward ("give" at the knees), the line of gravity is back of the hips, the knees are forward, straining the front thigh muscles, body weight is carried toward the balls of the foot, stretching the calf muscles and tending toward strain of the metatarsal arches of the foot, e) because of imbalance at pivotal points, there is muscle strain and stretching of the rear neck muscles, upper back muscles and of Ilio-psoas muscles, front thigh muscles, calf muscles and metatarsal arch sole muscles; and there is relaxation of front neck muscles, of abdominal muscles and of hamstring muscles; f) the shoulders are carried back of the sacral perpendicular line. A variation in the gorilla type may be encountered, in which the features are as here outlined, except that the pelvis is rolled forward downward.

The inclination of the pelvis may be considered the anatomical key to the variations in antero-posterior posture.

4. Distinguishing features of *functional lateral curvature* of the *spine* (Fig. 51) are (viewed from the rear): a) the line of gravity of the body, if projected upward from the mid position between the heels, passes upward in such a way that the parts above are displaced laterally in assymmetric positions; b) one knee is apt to be slightly bent (a habit, not necessarily a structural defect); c) the pelvis is tilted laterally downward; the ribs-pelvis curve is more marked on the side of the trunk opposite to the low side of the

FAULTY POSTURE—DEVIATIONS OF SPINE 171

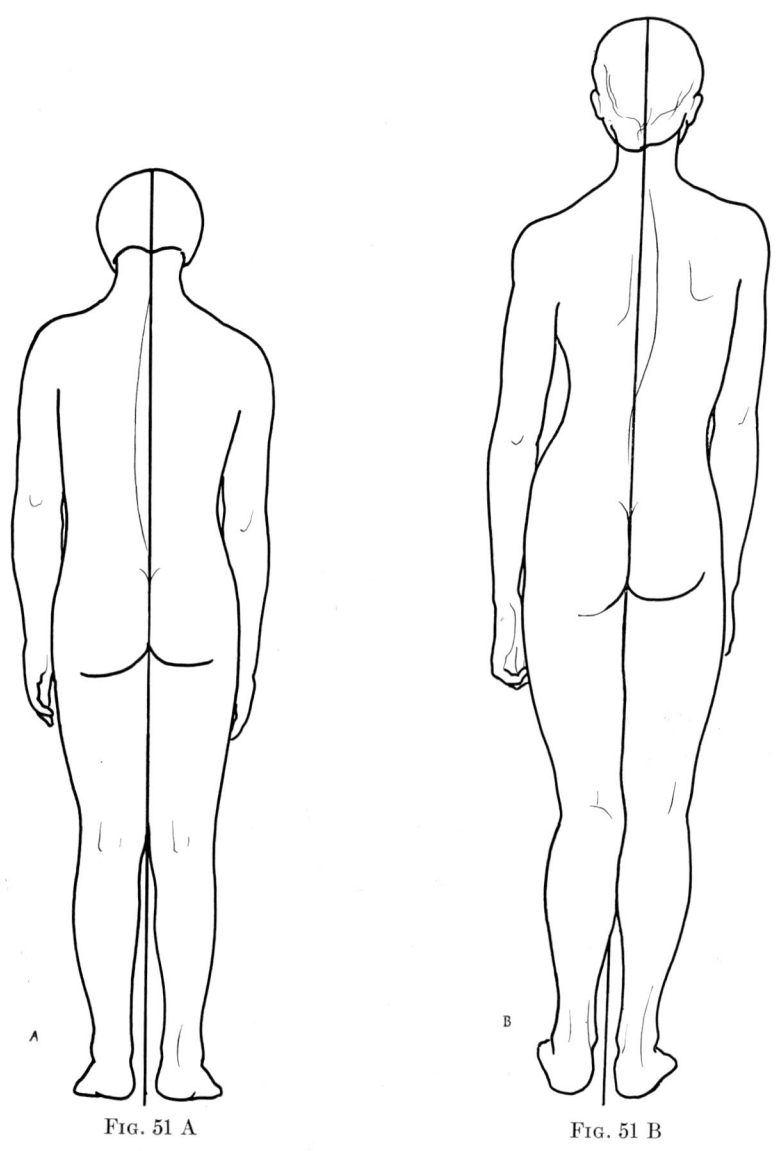

Fig. 51 A

Fig. 51 B

Fig. 51A. Functional lateral curvature of spine, C-type, rear, $\frac{1}{10}$ life
Fig. 51B. Functional lateral curvature of spine, S-type, rear, $\frac{1}{10}$ life

pelvis; d) the lines formed by tracing the spinal processes is curved —in the C-type, to the side of the low pelvis, in the S-type, to the side opposite to the low pelvis in the upper spine and to the same side of the low pelvis in the lower[2] spine; e) the shoulders are not carried level. They are tilted downward, in the C-type, to the side opposite to the low pelvis and to the same side as the low pelvis, in the S-type; f) there is little or no demonstration of rotation of the ribs with bulging of the rear of the ribs on one side, which latter is a characteristic of Rotary Lateral or Structural Scoliosis; and, g) finally, in functional lateral curvature of the spine, the lateral deviations cease in the prone lying position, or may be practically eliminated by body adjustments by the examiner.

Antero-posterior and functional lateral deviations of the spine are often associated in the same individual.

The writer's experience (Autumn, 1942), in the orthopedic examination of some thousands of Selectees for the Army, has led to the finding of an almost universal presence of functional weak feet in those who present functional lateral or functional antero-posterior deviation of the spine.

MEASUREMENT AND RECORDING

Record, in the standing position, should be made of I—the antero-posterior deviations from the normal and II—of the lateral deviations from the normal.

I. Record of the antero-posterior deviations from the normal include
    a. The relationship of the rear spine, especially of three points on the spine—seventh cervical vertebra, mid thoracic curve and mid lumbar curve—, to the sacral perpendicular;
    b. The amount of antero-posterior curvature of the two physiological curves—the thoracic and the lumbar;
    c. The inclination of the pelvis.

The relationship of the rear of the spine to the sacral perpendicular may be roughly noted by carrying a plumb-line near the rear of the spine; the plumb-weight being placed in the buttocks groove and the relationships to the string being noted. It is not so informative as is the method described below. (Note: Inclinations

---

[2] Note: The S-curve is usually a development from the C-curve, and leads, if uncorrected, into structural scoliosis.

FAULTY POSTURE—DEVIATIONS OF SPINE 173

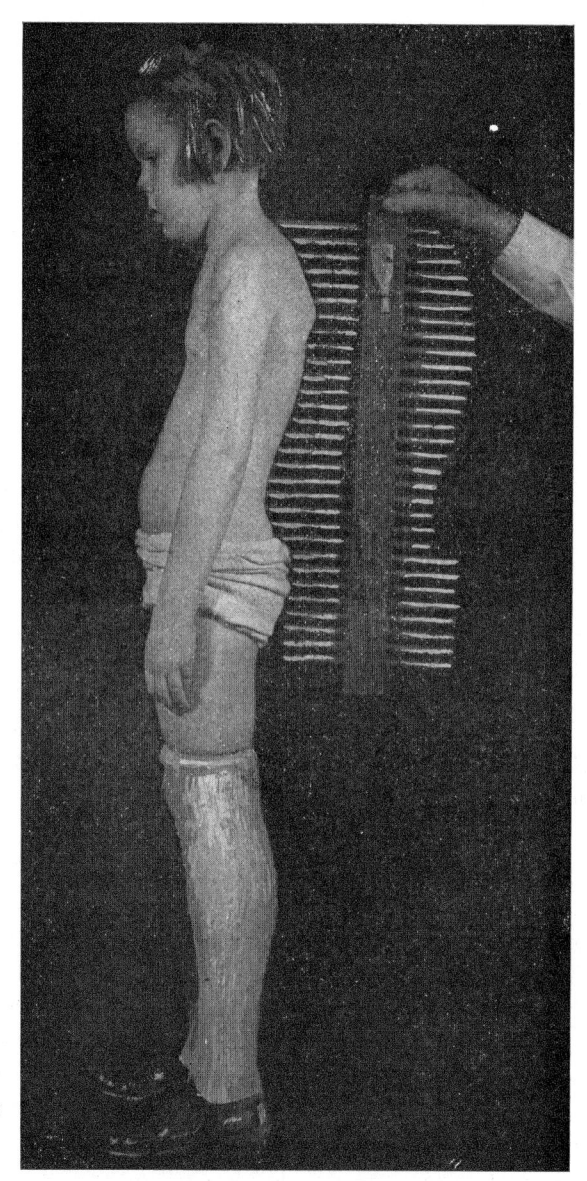

Fig. 52. Elkington's Spinometer, pendant and fitted to back

of the head and "give" at the knees or knee-back are observed but not measured.)

Apparatus necessary for recording antero-posterior deviations include

  a. Elkington's spinometer (Philadelphia) (Fig. 52);
  b. special recording paper;
  c. pelvic inclinometer (any surgical supply house);
  a. The spinometer consists of:
    1. two blocks of wood, each 57 cm x 4½ cm x 2 cm, placed and fixed face to face in such a way as to support sliding bars which are carried in circular grooves at 1.7 cm apart; and
    2. thirty sliding bars of wood, accurately measured to the same length (20½ cm) and each of about 1.7 cm diameter (like long unpointed pencils).
  b. The recording paper is of a stock type, checkerboard-ruled to 1 cm spacing.
  c. The pelvic inclinometer is of the type used by obstetricians.

Method of Measurement. The patient stands in habitual posture with bared back and buttocks. The spinometer is placed near the rear of the back and must hang perpendicularly from the examiner's hand (Fig. 52). With the other hand, the examiner gently pushes all sliding rods (from that opposite the seventh cervical vertebra to that opposite the first sacral segment) until each rod touches the examinee's back. Carrying it carefully, the apparatus is placed on its side, on the recording paper, in such a way that the supporting blocks of the spinometer are on (or parallel to) one of the perpendiculars found on the recording paper, and he also so places the spinometer on the paper as to insure that the lowest rod used (that recording the position of the first sacral segment) shall be on a printed perpendicular on the paper. This is important because this printed perpendicular becomes the "spinal perpendicular" of the permanent record. Before removing the spinometer from the paper, pencil marks are made, on the paper, of consecutive ends of each sliding rod which has been used in the measurement.

RECORDING

The spinometer is removed from the paper and recording is made from the dotted lines on the paper.

These paper records may be kept on file; or the examiner may measure and record the graph as follows (for example):

For a Kangaroo posture   0 : 0
                                                     21. :+5.4
                                                     41.5:+1.7
                                                     55. :+6.1
or for a Gorilla posture   0 : 0
                                                     11.2:+3.
                                                     41.2:−2.7
                                                     52. :+1.

Referring to the gorilla type record (just instanced), this means: that the first sacral segment is the fixed point for taking the "sacral perpendicular" and that therefore the record for this point is 0:0; that the point of greatest lumbar deviation is 11.2 cm up (on the sacral perpendicular) and is 3. cm in front; that the point of greatest thoracic deviation is 41.2 cm up and is 2.7 back of the sacral perpendicular; and that the seventh cervical spine is situated 52. cm up and 1 cm in front of the perpendicular.

Also, the arcs of the two physiological curves—the thoracic and the lumbar—may be measured on the graph paper and recorded thus:

1. For the lumbar deviation, measure the distance from the point marking the first sacral segment to the point where the graphed curve crosses the sacral perpendicular (e.g. 27.2 cm) and take the distance, already noted, of the lumbar curve from this chord (e.g. 3. cm). The record would read 3. divided by 27.2, equalling .1102, or the lumbar deviation is about eleven percent.

2. For the thoracic deviation, measure the distance from the graph crossing to the point marking to the seventh cervical vertebra (e.g. 25.7 cm) and the distance, already noted, of the dorsal curve from this chord (e.g. 2.7 cm). The record would read 2.7 divided by 25.7, equalling .1050, or a dorsal deviation of ten and one-half percent. These ratio records are helpful, but not as valuable as are the records of the graphic locations of the four points of the curve, noted above.

The pelvic inclinometer is an instrument consisting of a curved metal piece and a moving arm. The curved portion is passed between the thighs of the standing patient and so placed that one end of its curve touches the first sacral segment and the other touches the top of the symphysis pubis. The imaginary line between these two points is that of the inclination of the pelvis. On the movable arm is a spirit level; and on one or the other segment is a fixed curved dial, marked off in degrees of a semi-circle. When the movable arm is shown to be level, the angle it forms with the part of the apparatus that has been placed against the person to be examined is read on the dial, and recorded as the inclination of the pelvis.

These two instruments and their use as here indicated will constitute all that is necessary to record present conditions and improvement or otherwise under treatment, if antero-posterior deviation only exists. If, however, there is also functional lateral curvature of the spine, the standing position measurements, recorded in the next chapter, should also be taken (p. 203).

Mention however should be made of *shadow-graph* recording and of *photograph* recording, both of which methods have their warm supporters. In *shadow-graphing*, a long room is used; the person to be examined, stripped to the waist, stands sideways to and near a wall on which is fixed a large sheet of Manila paper. At the other end of the room is a single electric light bulb. The room is otherwise darkened and the examiner marks the shadow on the paper—the head, the neck, the back, the buttocks, the chest, the abdomen. For *photographic recording*, a similar distance is taken, from the camera to the examinee in side view. But in this case a frame, marked off with equidistantly placed horizontal and perpendicular cords, is placed near the examinee, and between him and the camera. The photograph records the outline checked off into many spaces. Both of these methods have the disadvantage of difficulty of precision, and of requiring space not always available; but they have the definite advantage of recording head and neck posture and abdominal and front chest contouring. Both the shadow-graph and the photograph may be used if there is functional lateral curvature. In this case, the person to be examined stands with his back to the light or to the camera.

Whatever method is decided upon, plea is here made that some mode of recording be used. Not otherwise can the examiner get a clear picture of what he is dealing with or of progress under treatment.

TREATMENT OF FAULTY POSTURE AND PLACE OF EXERCISE IN TREATMENT

The treatment of Faulty Posture consists in the proper application and apportionment of the principles of:

1. Balance of the structures including and those below the pelvis;

2. Temporary trunk support by the use of special corsets or body braces;

3. Muscle building and posture training exercises; and

4. The correction of harmful habits of hygiene—breathing, sleep, food, clothing.

1. Inasmuch as posture is largely a function of the standing position, it is obvious that the *pelvis* and the *segments below the pelvis* must be *balanced*. Any part of the basal mechanism which is in imbalance—which allows the pelvis to tilt laterally or more than normally antero-posteriorly—will inevitably retard and even prevent good balance of head, of neck and of carriage of shoulders and of trunk. Thus, weak feet saggings must be overcome (see p. 148). In the Kangaroo type of posture, knee-back is a frequent physical finding, and may be lessened by ordering a slight additional raise to the heels of the shoes. Or when functional lateral curvature is present, the temporary use of a raised shoe on the side of lowered pelvis, is an aid. The forward "give" at the knees, in the Gorilla type, is eased by lowering the heels of the shoes. Also any deformities of the ankles, legs, knees, thighs or hip joints should be compensated for by operation or by bracing. Such deformities may be those due to flexion contractions of the major joints or to bone distortions or shortening following rickets, tuberculosis, poliomyelitis or old fractures or dislocations.

2. The question of artificial *support* is an important one. Where faulty posture is due principally to muscle insufficiency, muscle development and posture training are most necessary. But, just

as in weak feet, in rotary lateral curvature of the spine and especially in poliomyelitis, the temporary use of artificial support is a real aid in attaining good posture. The boned corset is the best form of support for posture cases, although the celluloid corset or spinal brace may be indicated. In the Kangaroo type of posture, the corset should lessen the exaggerated forward roll of the pelvis and should carry the lumbar spine, the low chest and the abdomen backwards. In the Gorilla type of posture, the corset should roll the pelvis forward and should carry the low dorsal spine and the low chest forward and should hold the abdomen in. Artificial support should not be allowed to take the place of exercise, but in the sluggish and inattentive, it helps to train the "posture sense", and to overcome complemental muscle and ligamentous contractions. Artificial support should be discarded as soon as muscle development is adequate and posture sense is sufficiently acquired to "take over the job".

3. But *muscle development* and *posture training* constitute the principal, certainly the most lasting, means we have for the correction of body posture. Both of these types of exercises, muscle development and posture training, are essential. Details are found below (see p. 180). Posture training is best given to individuals alone, but a skilled instructor may group five or six persons, of the same sex and about the same age, in classes.

4. *Faulty habits* of *hygiene* should be looked to. The presence of adenoids and enlarged tonsils has a definite effect upon posture. Lessened intake of air in breathing leads to flat chest and poor posture. The removal of such, by operation, is often followed by great improvement in carriage. Inadequacies in eye or ear function leads to forward leanings of the head, neck and trunk or, if one sided, to twistings, and thus affect posture. They should be cared for. A patient may habitually sit in a chair or sleep at night in a position inimical to good posture. These should be discovered and corrected where possible. The custom of the Eighteenth Century of requiring a growing girl to sit for an hour a day strapped at waist and shoulders to a straight-backed chair, may seem harsh to us today, but it has its good points. Clothing may so constrict the shoulders and trunk as to prevent free growth in

the pre-adolescent and adolescent ages. For instance, a young girl whose breasts are developing rapidly may continue to wear a dress which was adequate for the childhood chest but which is not "full" enough to meet the breast growth, and the tightened dress will pull the shoulders forward. Also the length of the clothing both in front and in the rear may not follow normal growth of the boy or girl and may thus be a deterrent to natural body development. In boys, the waist belt is preferable to the suspenders to hold up trousers, because suspenders tend to pull the shoulders forward. There should be a well-balanced diet, the details of which must be individual matters. It is in order, however, to note that the extremes of weight development are factors in delaying the attainment of good posture. In the obese, the redundant fat is carried in such positions as to increase weight drags and thus to increase the curves of the spine. It especially tends to produce the Gorilla posture described (see p. 169). In pregnant women, the increasing weight carried in the abdomen, is a potent factor in producing, at least temporarily, the Gorilla type of posture. It should be met by special corsetting designed to carry the foetus upward and so lessen the forward carriage of the abdomen. Although normal activity is usually beneficial at this time, the specific exercises herein presented, should be omitted during most of the period of pregnancy. Corsetting, to keep the body-weight as near to the line of gravity as possible is the best means for maintaining posture in pregnancy. Excessive thinness increases the purely mechanical leverages of the spine. If this is accompanied by muscle insufficiency, faulty posture may result.

In discussing treatment of faulty posture, it is necessary to point out the *distinction* between poor posture due primarily to muscle weakness and faulty habits, on the one hand, and that due to disease or injury to the spinal bones. In faulty posture due to muscle insufficiency, the curves of the spine, although exaggerated, present *even curving lines* (Figs. 50b and c) and although the patient may complain of being "easily tired" there is rarely true pain in the spine. Where disease or injury to the bones is the cause of poor posture, the recorded curves are *angular* (at the level of the disease to injury) due to "crumpling in" of the bodies of one or more vertebra; and pain, especially pain on taking exercise, is a prominent symptom. The skilled physician has other tests for disease or injury to the spine, but the character of the curve and pain are the two outstanding factors in the differential diagnosis. It is necessary to note this distinction (especially by the non-medically trained worker), because the treatment of the two conditions is radically different. If faulty posture is due primarily to muscle weakness, corrective exercise is a main indication; temporary support may or may not be used as an aid. In faulty posture due to disease or injury of the bones, exercise to the part is positively harmful; support by artificial means (the details of which are not within the scope of this manual) is essential.

PREVENTIVE MEASURES IN RELATION TO POSTURE

1. Healthy babies should be allowed twenty to thirty minutes of unrestricted nakedness (in a well heated room) in their cribs, and encouraged or helped to roll and move arms and legs freely.

2. Pre-school and school children should wear unbinding clothing and should be allowed ever-increasing time for healthful play, out-of-doors.

3. All children from "first steps" onward should wear ample, firm-soled shoes.

4. Adults should walk more than is usual and practice good carriage while doing so.

5. Posture deforming factors, such as faulty breathing, inefficient sight or hearing, leg deformities from disease or accident, should be cared for—overcome or compensated for.

EXERCISES FOR FAULTY POSTURE

These exercises are divided into: A—Muscle Developing and B—Posture Training. In each group, many of the exercises presented tend to accomplish the results sought for in the other. But each end aimed at is necessary and best results are usually attained by using them in the order named.

A. *Muscle Building Exercises*

Ex. 1. Straight-trunk kneeling (see Fig. 26, p. 127). Movement—With hands on hips, bend trunk forward until head touches floor (Fig. 53). Progression may be made with "neck firm" position of hands and with arms stretched over head.

1. The knees flex further, with eccentric contraction of the Quadriceps muscles; the pelvis sways backward and the trunk bends forward with eccentric contraction of Glutei muscles, and the trunk continues to bend forward until head touches floor, with eccentric contraction of the Erector spinae and Latissimus dorsi muscles.

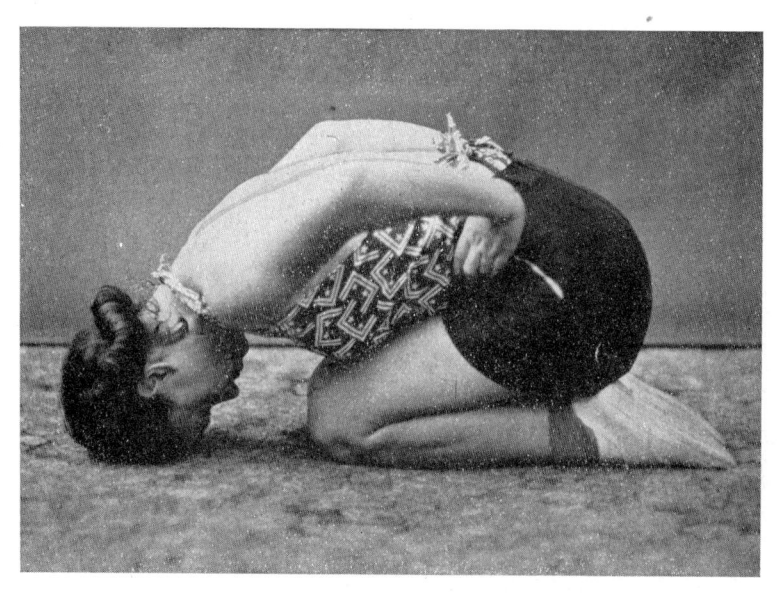

FIG. 53. Ex. 1—Kneeling, trunk bending forward

2. With the return to the starting position, head, neck and trunk are lifted, with concentric contraction of the Erector spinae and other back muscles; shoulders are drawn well back, with concentric contraction of the Trapezii and Latissimus dorsi muscles; the abdomen is drawn in, using all abdominal muscles; the thighs are extended with concentric contraction of Glutei muscles, and knees are half-extended to the starting position, with concentric contraction of the Quadriceps muscles.

Ex. 2. Straight trunk half-kneeling (see Fig. 27, p. 128). As in straight trunk kneeling except that one foot rests forward on the floor, the knee at right angles. Movement—With hands on hips, bend trunk backward (Fig. 54). Bend at hip joints only, keeping the trunk and head relatively in the starting position. Instruct to hold the abdomen in and not to bend beyond thirty degrees from the perpendicular. Progression may be made with "neck firm" and with arms over head.

FIG. 54. Ex. 2—Half-kneeling, trunk bending backward

The supporting segments, both thighs, legs and feet, are immoveable throughout the exercise.

1. The pelvis rolls backward on the hip joints, with eccentric action of the Ilio-psoas and other hip-flexor muscles, the trunk does not change its relationship to the pelvis except slightly extending in the low lumbar joints, with eccentric contraction

of the abdominal, spine and neck flexing muscles. (Note: Extension in the spine proper is to be minimized in this exercise as increase in lordosis is to be avoided.)

2. The return to the starting position is caused by concentric contraction of Ilio-psoas and hip flexor and of abdominal muscles.

Ex. 3. On hands and knees (Fig. 28, p. 129). Movement— Alternate thigh-leg-foot stretching backward to horizontal.

The thigh is extended, with concentric contraction of the Glutei and other Extensor muscles of the hips; the knee is in full extension with but little muscular action as the crucial and other knee ligaments mechanically hold the leg; for good form, the foot is in plantar extension, using the calf muscles. Also, to prevent the pelvis from rolling downward forward, the abdominal muscles are in tonic contraction.

Where gymnasium equipment is available,

Ex. 4. Opposite sitting. Sitting on bench, with the feet "hooked" in the stall bars. (See Fig. 32, p. 133.) Movement— With hands on hips, bend trunk backward (Fig. 55). Bend at first to thirty degrees from perpendicular trunk, gradually increasing, with increasing abdominal muscle strength, to head touching floor behind.

The thighs, extended knees, legs and feet, remain unchanged during the exercise.

1. The pelvis rolls backward downward, with eccentric contraction of Ilio-psoas and other hip flexor muscles; the lumbar spine extends backward with eccentric contraction of the abdominal and spine flexor muscles.

2. In the return to the starting position, the pelvis rolls upward and forward, with concentric contraction of the Ilio-psoas and other hip flexor muscles, and the spine returns to the starting position with concentric contraction of the abdominal muscles. (Note: Although this exercise tends to increase lordosis, it is such an efficient abdominal muscle developer as to be permitted; but it should immediately be followed by some spine flexor exercise.)

184 BODY POISE

Ex. 5, 6, 7. Lying face downward on the floor (see Fig. 30, p. 131).

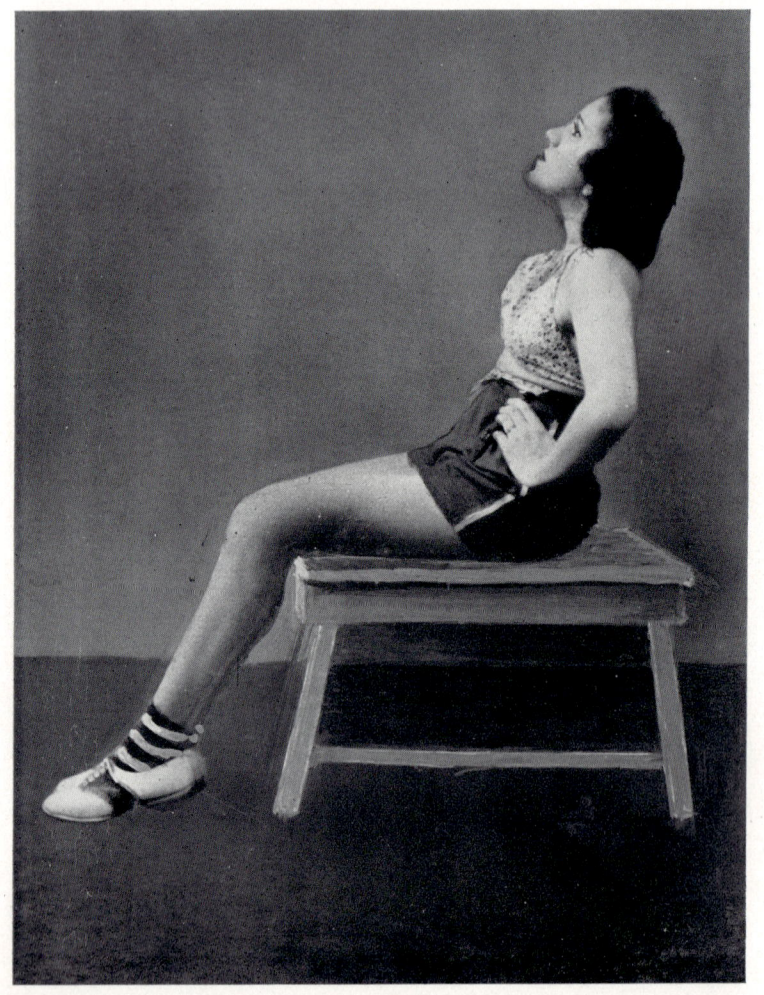

Fig. 55. Ex. 3—Opposite sitting, trunk bending backward (This model did not bend far enough).

Ex. 5. With hands clasped low behind the back, raise head and shoulders (*not* low trunk) (Fig. 56).

It is important that this exercise shall be limited to upper thoracic spine, neck, head, shoulders and arms. From about thoracic vertebra VI and upward through the upper thoracic, the cervical vertebra and the head, there is extension of the spine, using the upper back and rear neck extensor muscles concentrically, and there is marked retraction of the shoulder girdle (adduction of the scapulae) and slightly increased adduction and extension of the humerus, using the Trapezii and the Latissimus dorsi muscles. (Note: This is a useful exercise to develop the muscles necessary to hold good posture of head and neck and of shoulders and to overcome round back.)

Ex. 6. Raise alternate straight thigh-leg-foot.

The mechanism and kinesiology is similar to that of Exercise 3 above.

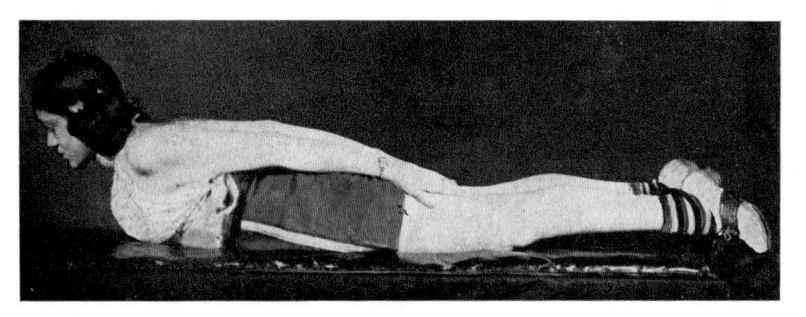

Fig. 56. Ex. 4—Prone-lying, head and shoulders raising—"The Seal"

Ex. 7. Raise alternate leg (bend at knee).

This is simple leg bending to a right angle at the knee, with eccentric contraction of the hamstring and calf muscles. It is of value in overcoming the knee-back deformity developed in the "kangaroo" type of faulty posture, and incidentally in some phases of poliomyelitis disability.

Ex. 8, 9. Lying on the back on the floor (see Fig. 29, p. 131).

Ex. 8. Movement—With knees drawn up, feet resting on the floor, bend both knees to the chest (Fig. 57).

The position of lying on the back relaxes much of the mechanism of the human frame and allows concentration in this exercise on the part to be developed. The starting position for this exercise, with the knees bent and the feet resting on the floor, is given to minimize possible harmful strain on the sacro-iliac joints and low lumbar spine while taking the exercise. The thighs (with knees flexed) are fully flexed until the knees touch the abdomen. This uses the Ilio-psoas and other hip flexor muscles, and is an important abdominal exercise because, as soon as the feet leave the floor, the weight of the limbs tends to roll the pelvis forward downward, and to prevent this there must be tonic contraction of the abdominal muscles.

Fig. 57. Ex. 5—Lying, double knee-bending to chest

Ex. 9. Movement—With stretched arms resting over head on the floor and legs straight ahead; arm flinging forward and trunk raising to sitting, to far reach forward, to fingers touching toes (Fig. 58).

In this starting position, the thighs and legs are extended and the arms are extended over head and resting on the floor. Also, to meet over-balanced weight of the moving segment, it may be necessary for the feet to be held by stall bars or other means. To get momentum, the exercise begins with arms flinging forward but is immediately followed by flexion of the lumbar spine which continues to flexion of the upper spine and with flexion of the pelvis on the thighs and may require slight flexion of the knees, to allow the finger touching to the toes. Although many other muscles are used, this is principally an abdominal muscle developer.

Ex. 10, 11, 12. Half-prone lying at end of table, feet resting on the floor (see Fig. 31, p. 132).

This is an excellent starting position, as it relaxes much of the human mechanism while allowing concentration on the head,

FIG. 58. Ex. 6—Lying, with arms flinging forward, trunk raising to sitting to far reach forward.

neck and the arms and shoulder girdle, or on the lower extremities.

Ex. 10. Movement—Alternate straight thigh-leg-foot raisings to horizontal (Fig. 59).

As a developer of low back and thigh extensor muscles, the kinesiology of this position is similar to that of Exercise 3 and Exercise 6; but with the longer range of motion (as compared with Exercise 6) and the greater leverage of the fully extended limb (as compared with Exercise 3), it is a stronger developer of these muscles.

Ex. 11. Movement—Double straight thighs-legs-feet raising (Fig. 60).

188   BODY POISE

Fig. 59. Ex. 7.—Half-prone lying, alternate thigh-leg-foot raising

FAULTY POSTURE—DEVIATIONS OF SPINE 189

Fig. 60. Ex. 8.—Half-prone lying, double thighs-legs-feet raising

This exercise adds much effort to the muscles involved in the previous one and requires hand grasps on the sides of the table to prevent over-balancing by the greatly increased weight of the moving part. This exercise is not to be given until much previous training has developed the muscles sufficiently to make it safe.

Ex. 12. Movement—With hands clasped low behind the back, raise head, neck, shoulders and upper trunk.

The trunk from the low chest downward, is not raised. Progression may be made by changing the arm position to arms stretched sideways, to arms over head.

The kinesiology of this exercise is similar to that of Exercise 5. It is important to limit it to the upper half of the thoracic and the cervical spine, head and shoulders and arms. The exercise is useful in developing the muscles necessary for good carriage of head, of neck and of shoulders and in overcoming Round back.

Ex. 13. Sitting on a bench, a one-pound dumbbell on the floor between the feet. Movement—Reach down and grasp the dumbbell with the right hand, carry it to shoulder to high over head, to shoulder, to floor; grasping the dumbbell with the left hand, carry it to shoulder, to high over head, to shoulder, to floor (eight counts).

Many muscles and many joint mechanisms are used in this exercise. It is given to "limber up" the spinal and alternate shoulder-girdle mechanism. As such, it uses back and shoulder-girdle muscles, in the development of good posture. It has but little effect upon the abdominal muscles.

Ex. 14, 15, 16, 17. High sitting (on a table) legs and feet hanging.

Ex. 14. Alternate leg raising (straighten at knee) Fig. (61).

For development of Quadriceps muscles.

Ex. 15. Alternate hip flexing (Fig. 62), then knee extending (Fig. 63) and return to starting position.

For development of hip flexor and Quadriceps muscles.

Ex. 16. Place right ankle over left knee, raise the foot and sole upward and the toes under (Fig. 47, p. 160).

FAULTY POSTURE—DEVIATIONS OF SPINE 191

Ex. 17. Place the left ankle over the right knee and turn the foot and sole upward and the toes under.

Fig. 61. Ex. 9—High sitting, alternate leg raising

The kinesiology of this is described under Foot Exercises (see p. 159).

It will be noted that, in the exercises for muscle building, for faulty posture, standing positions are avoided. This choice is

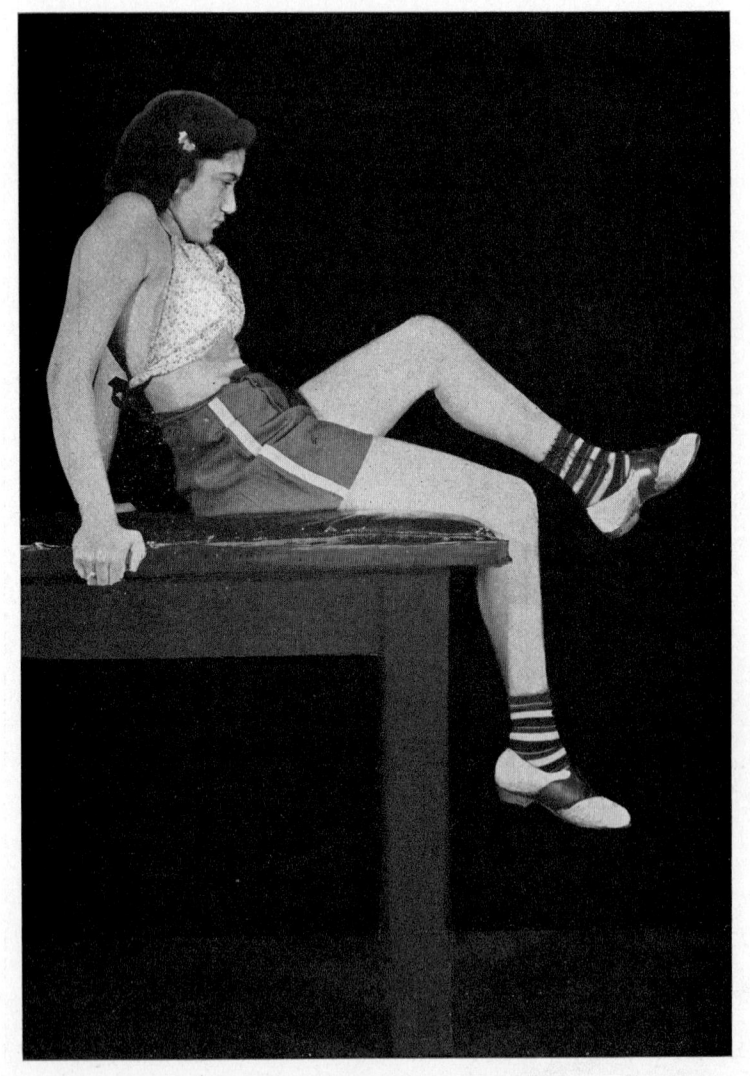

Fig. 62. Ex. 10—High sitting, alternate hip-knee bending

made because in the kneelings, on hands and knees, the lyings and sittings, concentration on the parts to be developed is better at-

FAULTY POSTURE—DEVIATIONS OF SPINE 193

Fig. 63. Ex. 11—High sitting, alternate straight limb raising

tained than in the standing positions. A considerable degree of muscle development will usually be necessary before the real posture training may be started.

B. Posture Training

Exercises for posture training are quite as essential as muscle building exercises. The standing positions are now used. A full length wall mirror is a material aid. The doctor or instructor must himself have a clear concept of what constitutes good posture. He should guide the one under his charge in changing his body alignment to attain and maintain good posture. The first five exercises (below) are designed to train the muscle sense in maintaining good posture, in head, neck and trunk, whilst various modifications in his balance are made by changing the mechanism of the lower and of the upper extremities.

Ex. 1, 2, 3. Standing at "attention".

Ex. 1. Movement—With hands on hips, heel raising, knee bending, knee stretch, heel sinking (Toe knee-bend standing) (Fig. 64).

Ex. 2. Change arms to hands behind the neck (neck firm) to arms over head (stretch standing).

Ex. 3. Combine exercises 1 and 2—that is, in toe-knee-bend standing, as a temporary starting position, change arms to various positions.

These exercises beging the element of Balance training, under unusual conditions. Muscle Co-ordination, in half completed leg positions, is drilled. The hips are quarter-bent, using especially the Glutei and to a lesser degree the hip flexors; the knees are half-bent, using especially the Quadriceps and, as steadiers, the hip flexors; the ankles are neither dorsi-flexed nor plantar-flexed, but the raised position of the heels requires tonic contracting of the calf muscles, and for steadiness many other foot-control muscles are called into play. The changes of arm positions, in the three exercises, are included further to develop the elements of balance and muscle co-ordination, in the trunk and legs, under progressively varying distractions in the mechanism of the arms. An alert instructor may add other arm changes, and leg changes, such as alternate limb straightenings and placements forward or sideways, when previous training warrants it.

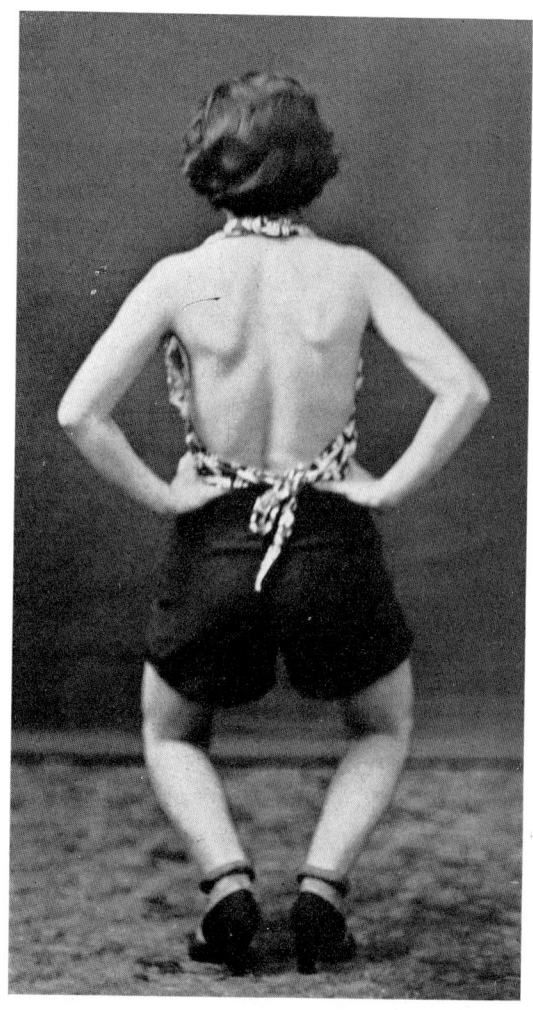

Fig. 64. Ex. 12—Wing toe-knee-bend standing

Ex. 4. Military marching.
Ex. 5. "High step" marching.
Also, much valuable posture training can be accomplished by

carrying a light, and progressively increasing, weight on the head whilst doing exercises 1, 2 and 3 (above) and, in Exercise 4 and Exercise 5, in walking away from and returning to the mirror-view for check-up.

Ex. 6—Dr. Eliza Mosher's posture training exercise for correction of Gorilla type with downward roll of the pelvis: (Note: This downward rolled pelvis is a frequently found variation in the more usually upward rolled pelvis, in the Gorilla type.) (Fig. 65a.) Starting position—Standing with knees straight, but with trunk bent forward as far as possible (back curved), head bent forward and arms hanging loosely (Fig. 65b). Movement—The movement is a gradual and progressive muscle control from below, upward, as the trunk is straightened to the position of military "attention". Thus the patient is instructed:

a) to make the knees firm and draw in the abdomen (Fig. 65b);

b) to begin to straighten the lower trunk (Fig. 65c);

c) to continue straightening the low trunk and to roll the hips (pelvis) rear downward (Fig. 65d) and front upward;

d) continue to draw in abdomen and continue straightening of the trunk to the base of the neck (Fig. 65e);

e) to draw the head and neck backward and the shoulders backward, until the military "attention" position is attained (Fig. 65f). (Five counts and repeat.)

(Note: This exercise should be taken sufficiently slowly and accurately to allow each phase to be attained and to be maintained while each succeeding position is added. At first it should take rather more than a minute to complete, and never should it be done rapidly. It involves much neuro-muscular training in co-ordination.

The Kinesiology of the Mosher exercise is:

Phase b—concentric contraction of Quadriceps and of abdominal muscles;

Phase c—partial extension of lower spine, using Erector spinae muscles concentrically;

Phase d—upward roll of pelvis, using Glutei muscles concentrically, and more contraction of Erector spinae muscles;

Phase e—further contraction of the back muscles and further attention to the abdominal muscles;

FAULTY POSTURE—DEVIATIONS OF SPINE 197

FIG. 65. Ex. 13—Eliza Mosher posture correcting exercise. (In six phases)

Phase f—final extension of the upper and cervical spine and good head carriage, with use of the upper spine extensors, and retraction of the shoulder-girdle (adduction of scapulae), using the Trapezii and Latissimus dorsi muscles.

The value of this series of phases is both pedigogical and physical. When properly done, it trains the individual from simpler to more and more comprehensive use of muscles, as it is essential that, as each new muscle group is used, those already employed must maintain their action. In particular, it emphasizes, to the individual, the upward rolled pelvis, the indrawn abdomen and the retracted shoulders.

Appreciation is here recorded of the excellent posture training attainable at military establishments, schools and camps, under able drill masters, and also where Swedish gymnastics are effectively used. Posture training is much more than muscle building. Principles of Pedagogy and of Psychology enter into its fulfillment.

SUMMARY

1. The normal or neutral and three faulty types of posture are defined, by noting the relative positions of principal segments of the body, in the standing position, to the line of gravity of the body, and by indicating the faulty inclination of the pelvis in each.

2. Exercises are classified as muscle-training and posture-training. The muscle-training exercises are given from starting positions other than the standing as more conducive to concentration on the parts involved. Posture-training is taken in the standing position as only thus will their indications be met.

3. Methods of measuring and recording are included to insure a clear picture of the condition cared for and to know whether the treatment is efficient or not.

4. The entire treatment of faulty posture is outlined. It consists of a clear understanding and use of the principles of:
   a) balance of the pelvis and segments below the pelvis;
   b) the temporary use of artificial support, preference being given to the boned corset;
   c) progressively graduated exercise—both developmental and postural; and
   d) the correction of harmful habits of posture.

5. Distinction is made between faulty posture due principally to muscle insufficiency and that due to bone disease or to fracture. This distinction is included in order to warn the non-medically trained reader of the radically opposite nature of treatment indicated.

Chapter 5

ROTARY LATERAL CURVATURE OF THE SPINE—
SCOLIOSIS

DEFINITION AND MECHANISM

Rotary lateral curvature of the spine is a condition in which the spine deviates sideways, with uneven shoulders and with asymmetrical distortion of the spine and bony frame-work of the thorax. Scoliosis usually starts as simple functional lateral deviation of the spine, with weakness of the muscles of the spine and of the abdomen, as in Poliomyelitis, although distinct "anatomical variation" is sometimes discovered, by X-ray examination, at or near the fifth lumbar vertebra and the sacral joints, or at the thoracico-lumbar junction, and this may be the primary causative factor. Rickets sometimes, but rarely, is the cause. Also, unequal balance of the structures below the hip bones may be encountered and must be considered as affecting the lateral spinal balance. A so-called "idiopathic" type, or one without known cause, has long been included in classifications of Scoliosis. But, with increasing study of the deformity, more and more of these have been placed in the list of known causes, and the word "idiopathic" is less and less used. However, whether as cause or effect, asymmetric muscle development is always a marked feature of the condition.

The deviation is at first simple, and with one curve (see Fig. 51a, p. 171) and can momentarily be corrected by suggestion or by slight manual pressure. This is the functional lateral curvature of the spine, considered in the preceding chapter. If uncorrected, in this stage, by proper muscle and posture training, it progresses to a double curve (see Fig. 51b, p. 171) (although occasionally a single curve persists) and finally spine and rib shapes are altered, and rotations of the spine and ribs occur and become more and more fixed. Statistics show that the type, Right Thoracic Left Lumbar, constitutes about eighty percent of all lateral curvatures

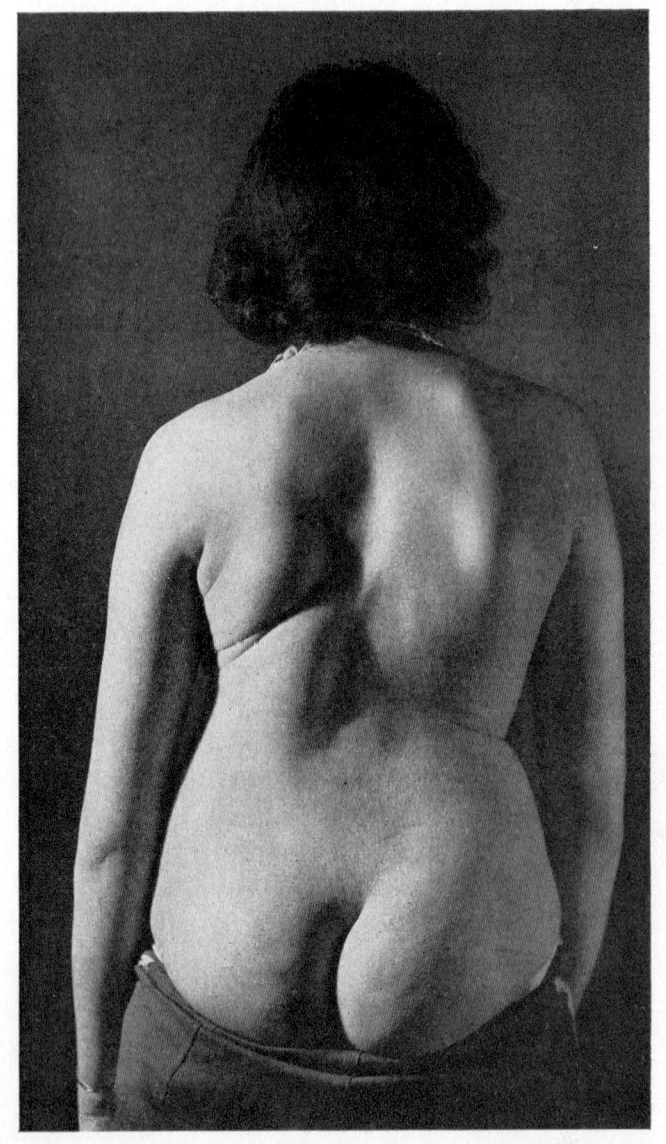

Fig. 66. Rotary lateral curvature of spine—Right thoracic left lumbar type

of the spine. It will be noted therefore that, for clarity in what follows, the Right Thoracic Left Lumbar type will be discussed here and that one dealing with the opposite type—Left Thoracic

Right Lumbar—must reverse all that is thus said, changing the words "right" to "left" and "left" to "right."

In Right Thoracic Left Lumbar Rotary lateral curvature of the spine, the characteristic features are (Fig. 66):

1. The curve of the spinous processes, in the thoracic region, is to the right and the curve of the spinous processes, in the lumbar region, is to the left;

2. The general trunk-lean is to the right of the sacral perpendicular;

3. The left shoulder is carried lower than the right shoulder;

4. There is a rearward bulge of the right rear chest wall, greatest at the level of the right thoracic lateral deviation and a forward flattening of the right chest wall. There is also a rearward flattening of the left upper chest wall and a forward bulging of the left lower front ribs. There is a rearward bulge of the left side of the lumbar spine, greatest at the level of the left lumbar lateral deviation. Both of these bulgings and secondary chest flattening are due to rotations of the spine in their respective regions—the thoracic spine and rear chest twisting to the right, the lumbar spine twisting to the left. Also the vertebrae are rotated in such a manner that their bodies are in even greater lateral deviation than are the more evident spinous processes. That is, we easily measure the deviations of the spinous processes, but (from X-ray film and examination of an anatomic specimen of a scoliotic trunk) we know that there exists greater deviations, vertebra for vertebra, in the bodies; (see Fig. 73a, p. 213).

5. There is usually exaggeration of the physiological anteroposterior curves of the spine.

MEASUREMENTS

For accurate knowledge of the condition and relationships of the bones involved—the spine and the ribs—the X-ray films should be used; but the writer's measurements[1] (discussed below) afford an excellent working basis for frequent comparisons and are easily and quickly made. An initial X-ray study of the deformities is the more valuable if taken in three positions of the patient, in each of which the patient is in a half-prone lying posi-

[1] Truslow: Am. Phys. Edn. Review, Vol. VI, No. 3.

202　　　　　　　　　　　BODY POISE

tion, the film being placed on the table beneath the patient's prone trunk. These positions are as follows:

1. Half-prone lying *without* trunk-twist (Fig. 67a);
2. Half-prone lying with as much upper-trunk shoulder twist to the *right* as possible (Fig. 67b);

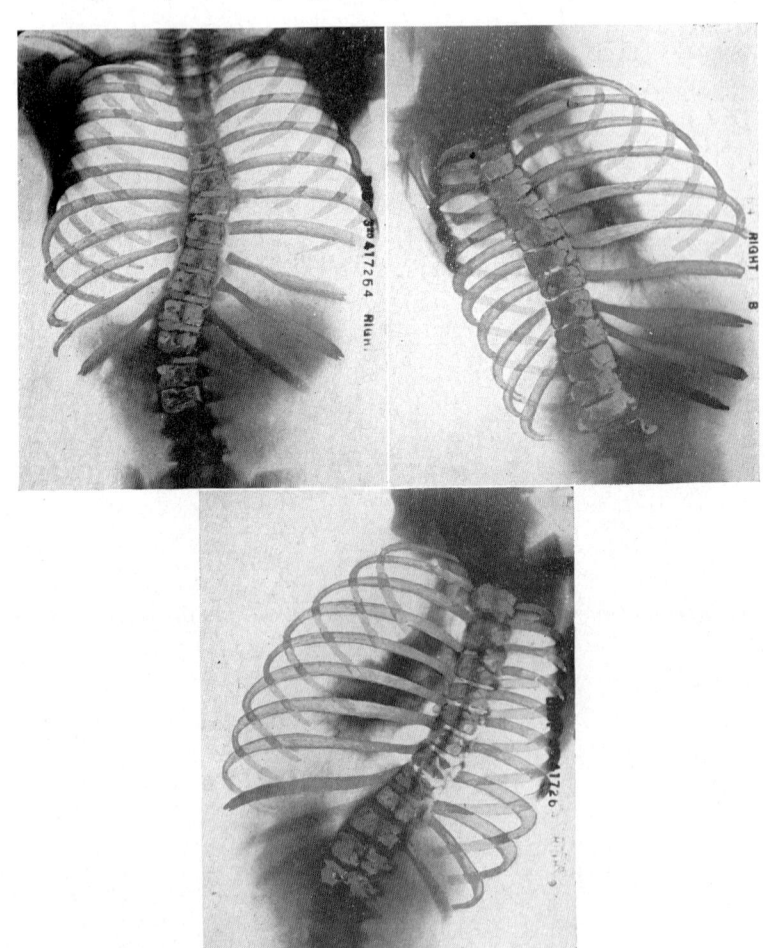

FIG. 67. Three X-rays of a scoliotic spine, taken with subject in prone-lying; that is, from rear forward (retouched).
 A. (upper left). Without gymnastic twist to the upper trunk.
 B. (upper right). With upper trunk-twist to the right.
 C. (lower). With upper trunk-twist to the left.

3. Half-prone lying with a twist to the *left* of the shoulders and upper trunk to as much as is possible (Fig. 67c).

If may be, each of these three pictures should include in one film the entire spine, from the seventh Cervical to the Sacrum, and the Ribs. With these sets of films, the examiner may get an excellent idea of the total deviations and of the rotations. (The argument for right upper trunk-twist, in treatment, is discussed below, see p. 211). The films do not afford a good record of the general trunk-lean nor of the relative carriage of the shoulders in the habitual vertical positions of the patient. But they are important in determining how much present correction of the deformity is attainable, by right upper trunk-twist, and in discovering certain anatomical variations which may be present in the lower Lumbar and in the Lumbo-sacral regions. It is important that a careful X-ray study, as here outlined, be made at the beginning of treatment for scoliosis.

However, X-rays are expensive and require expert technique. As a working basis for information as to progress under treatment, certain measurements have been found of value. They can be taken and recorded in from ten to fifteen minutes. They are made with the patient in the habitual standing position and in the prone-lying position.

A. *Measurement* in the *standing position*. The patient stands with the back exposed from the base of the neck to the buttocks' fold (Fig. 68). A strip of adhesive plaster is placed on the spinous processes from the seventh Cervical to the first Sacral (top of the buttocks' fold); the successive spinous processes are felt and marked on the adhesive plaster; the level of the right and left scapular inferior angles are projected and marked on the margins of the adhesive plaster; a plumb line is then hung in such a manner that its weight will be carried in the buttocks' fold and the upper end of the cord to the right or left (as may be) of the seventh Cervical marking (Fig. 69). Where the cord passes the level of the seventh Cervical point, a mark is made on the adhesive plaster. To make a permanent record of this graph, the adhesive plaster is transferred from the patient's back to any flat surface and the following lines drawn and distances measured. A line is drawn with a ruler guide, from the seventh Cervical to the first Sacral dot. It is called "spinal height." A line is projected from it to the dot representing greatest thoracic deviation; another to

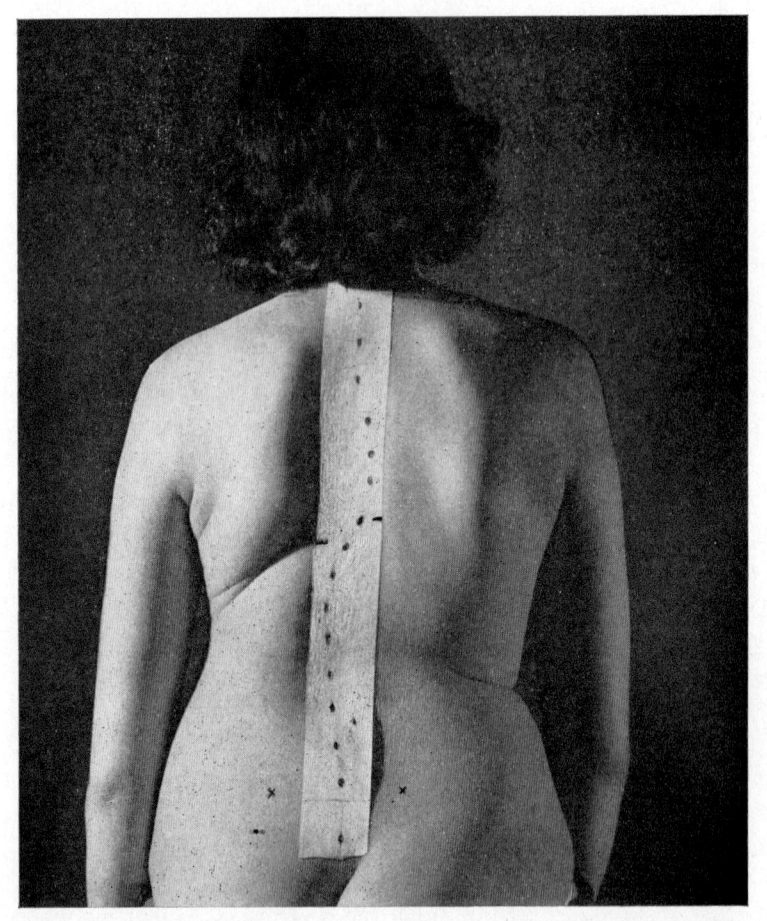

Fig. 68. Measurement of lateral curvature of spine, standing. Adhesive plaster strip placed and marked, to record lateral deviation and relative height of scapulae.

greatest lumber deviation; both markings for the scapular angles are projected across the adhesive strip to get their relative levels, each to the other. Measurements are taken as follows:

a—of thoracic deviation;
b—of lumbar deviation;

c—of spinal height;
d—of relative scapular levels;
e—of projection, to the right or left, of seventh Cervical vertebra to the Sacral perpendicular.

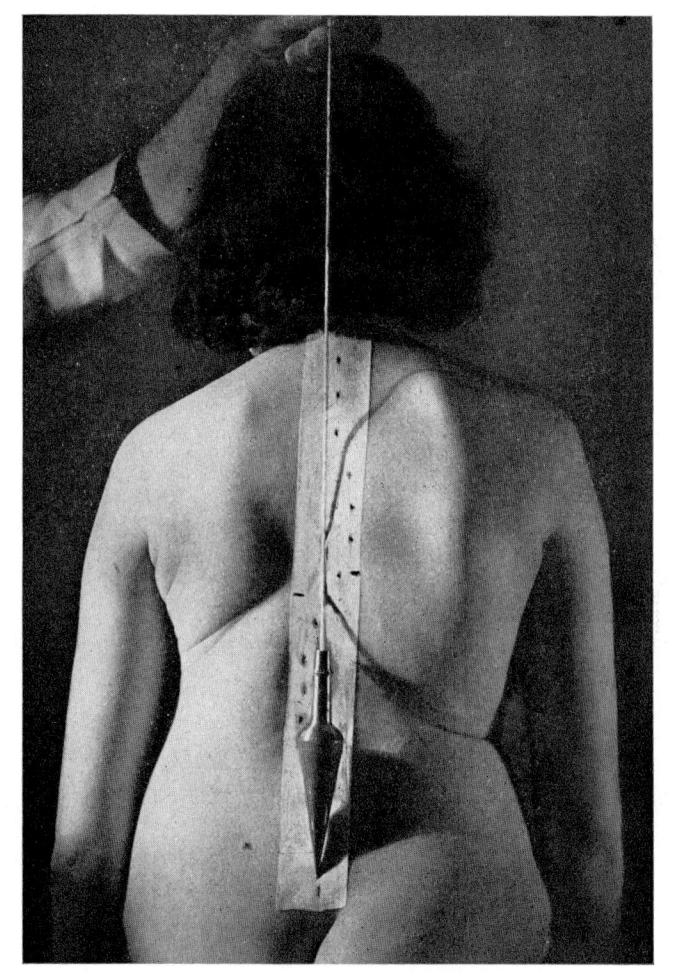

Fig. 69. Measurement of scoliotic spine, to record relation to sacral perpendicular

A history sheet record of the above reads, for example:

"Spine standing — Spinal deviation $\dfrac{3.2\ +\ 1.5-}{45.}$ = .1044 or $10\tfrac{3}{8}$ percent.

Carries left shoulder 3.3 lower.
Carries seventh Cervical 1.6 to right."

The above is the record of the elements of deformity in standing or weight-bearing posture.

B. To obtain a further record of bony changes, including rotation, and especially of the present best correction of the deformity, weight-bearing must be eliminated. The patient is placed in a

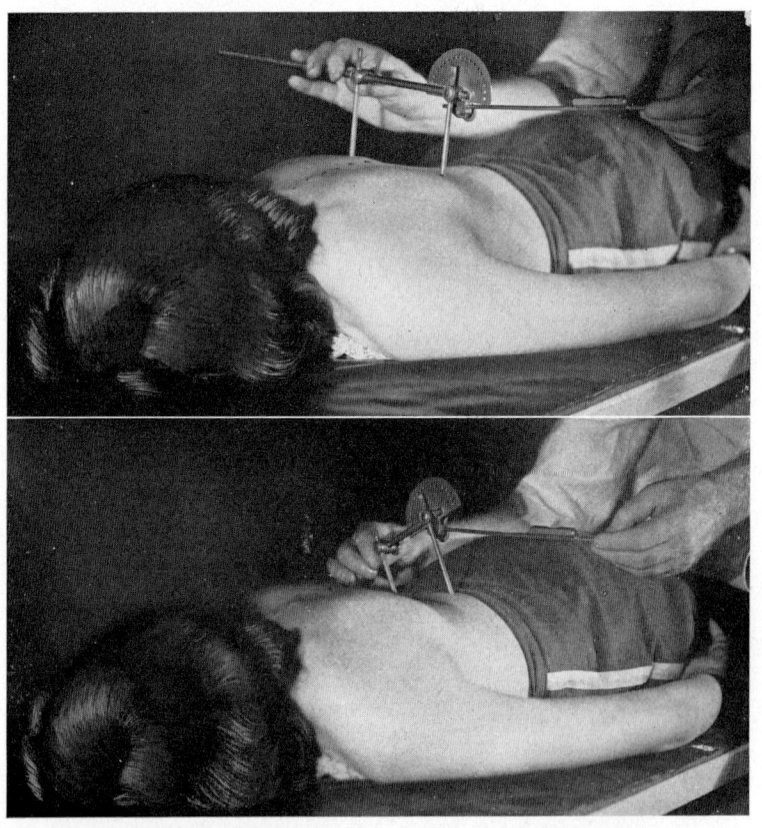

FIG. 70 (upper). Measuring rotation in scoliosis—thoracic
FIG. 71 (lower). Measuring rotation in scoliosis—lumbar

standard *prone position* on the examining table. The best correction possible is made by the examiner. Another strip of adhesive plaster is used upon the exposed back from the seventh cervical to the buttocks' fold. Successive spinous processes are felt and marked. Rotations in degrees are obtained by the use of a rotometer (Figs. 70 and 71), having two movable arms, one of which carries a spirit level and is held beside the patient's trunk, while the other arm is placed across the patient's back at the levels of greatest lateral deviations, which are always also the levels of greatest rotation deformities in the spine. At the joint of these two arms of the rotometer, are fixed (to one of them) a semi-circular disc marked off in degrees; (to the other) a small indicator which will record on the disc the angle which the two arms make each to the other when placed as noted above. The greatest thoracic rotation is first noted and recorded (Fig. 70); then the greatest lumbar rotation (Fig. 71). The adhesive plaster is transferred, ruled and measured for spinal height and for thoracic and for lumbar deviations, as when taking these measurements with the patient in the standing position. The history sheet record (of the same patient as noted above) might read, for example:

"Spine prone — Spinal deviation $\dfrac{2.2 + 1.1 -}{45.}$ = .0733 or $7\frac{1}{3}$ percent.

Rotations, 8 degrees and 5 degrees."

Comparison of the relative measurements of the standing and of the prone lying positions is an aid to prognosis. And comparison of succeeding measurements is particularly important in determining how effective is the treatment, and what feature or features of deformity correction must be emphasized in continuing treatment.

THE TREATMENT OF SCOLIOSIS

The treatment of lateral curvature of the spine consists in a proper application and apportionment of the principles of:

1. Balance of the structures below the pelvis;
2. More rapid correction of the deformities with plaster of Paris jackets and support of the corrections attained by the use of corsets or braces;
3. Corrective, developmental and posture training exercises; and occasionally
4. Permanent fixation of corrections attained (especially where *paralysis* of muscles is the underlying cause) by "internal splinting," i.e., by fusion or bone grafting operations.

1. *Balance of Structures below the pelvis* is important. If the pelvis itself is not level laterally, it would be futile to carry on treatment for the spine itself and yet allow an habitually inclined pelvis to continue the potentiality of the deformity. The laterally inclined pelvis may be due to unilateral congenital dislocation of the hip; to unequal length of the lower extremities, from thigh or leg or foot fractures, or delayed growth following infantile paralysis or old hip joint disease; to unilateral knock knees or low legs, or to unequal flat foot. The recognition of any of these conditions is part of the routine of the orthopedic surgeon's examination; the recognition or suspicion of them by others should lead to a careful examination by the expert. Where the correction of the conditions themselves is possible, it should be made. When this is not possible, or is refused by the patient, compensation for a shortened limb should be made by a cord sole and heel beneath the shoe, to raise the shoe sufficiently to level the pelvis laterally.

2. Much time may be saved to the patient by the preliminary use of rapidly correcting *plaster of Paris jackets*. Details of this will not be dealt with at this time. Effective measures of procedure are well known. Each surgeon will use that which he knows best. But some features of technique are worth emphasizing. First, one must have a clear idea of the elements of the deformity to be corrected. The writer finds it more effective to depend upon the application of the plaster jacket to correct the upper trunk-lean and the low shoulder; and upon the exact placing of subsequent paddings to lessen the spinal deviations and the rotations; rather than to emphasize the correction of all deformity elements at once by the position of the patient upon whom the jacket is built. Negative windows are cut out and first paddings to correct spinal deviation and rotation are applied a few days after the jacket is well set. Succeeding passings are applied once a week for four to six weeks after the application of the plaster jacket. During that first jacket period, the rigidity of the jacket itself prevents any further correction of the faulty upper trunk-lean and in the low shoulder; but much correction of the spinal deviations, of rotation and of anterior rib deformities

may be obtained. This point is emphasized as it is believed that hazy understanding of it accounts for indifferent success.

A second plaster jacket, with further improvement of the deforming element and with four to six weeks of paddings, at one to two weeks intervals, may well be undertaken. At the end of three months of this, the figures representing the spinal deviation (see p. 205) should have been reduced about one-half and that representing spinal height should have been slightly increased. The upper trunk-lean, as represented by the relation of the seventh cervical vertebra to the sacral perpendicular, should have been carried nearly to or perhaps passed across the certical line; and the shoulders should have been leveled. During this time it is usually possible to reduce the figures indicating rotation about one-half. Although the correcting forces have been gradually but steadily applied, although the cooler months of the year may have been chosen and the "scratcher" faithfully used, the patient's skin and often the patient's disposition will not tolerate more than three months of these jackets at one time.

Plaster of Paris jackets are also used, for much more rapid and forcible correction of the deformity, by employing segmented jackets, controlled by hinges and turn buckle screws. Lovett[2] initiated this and Butte[3] and others have amplified by inclusion of the neck and rear of the head and of one thigh to insure better fixation of the segments to be corrected. In applying the plaster jacket, incorporation of a hinge on the convex side and of a turn-buckle apparatus on the concave side of the deformity is made. When the plaster is thoroughly hardened, horizontal circumferential cutting is made at the level of the hinge. This is carried entirely around the trunk and thus divides the entire plaster into controllable segments. The turn-buckle screw is used to open the segments gradually. This correction is given first to the "primary" curve, and then (sometimes) to the "secondary" curve. To determine which is the primary curve, Butte[3] states that, at the original examination, the primary curve is found more fixed

[2] Lovett, Robert W.

[3] Butte, F. L., Scoliosis Treated by the Wedging Jacket. Jl. Bone & Joint Surgery, Jan., 1938.

on lateral bending and the secondary curve less fixed on side bending.

Steindler[4] uses the plaster jacket to get a new balance of the spine, by the attainment of "contra-lateral curves" above and below the original one. He states that the compensation sought "consists in so establishing counter curves as to eliminate overhang and to bring the body-carriage over the pelvis." This usually requires corrective plaster of Paris jackets, often followed by an internal splinting operation, but in mild cases it may be accomplished by corrective gymnastics.

After much correction of the deformity has been attained by using plaster jackets, what is now to be done? We know that some of the deformity will recur if we do not hold what we have attained and so train the muscles, by intensive exercises, that they will increasingly be able to assume the task of natural support with ever lessening artificial support. (See diagram, p. 6.) The *brace* or *corset*, planned for at the time of change from the first to the second corrective plaster jackets, should now be ready.

A truly retentive brace is difficult to attain, but is important. The requisites are:

a. Ability to hold the correction attained;

b. "Fool-proof"—the patient must be able to apply it with reasonable accuracy;

c. Extensibility to meet normal growth and longitudinal spine lengthening and lateral compressibility to follow further deformity correction which proper intensive exercises should surely give; and

d. Finally, if possible, self correction.

The carefully molded celluloid corset is the lightest and most easily worn, and the surest way to meet the first two needs; the Knight spinal brace can be modified to meet also the third need and is fairly efficient. The writer believes that a brace meeting the fourth need—further self-correction on the stop-joint principle—should be possible but not yet attained. The brace or corset must be worn by night as well as by day at first. Through all of the corrective jackets and retentive brace period, a school

[4] Steindler, Arthur, Text Book of Surgery (Christopher), Saunders, Phila., 1939.

girl or a working girl may usually continue at her duties, except perhaps for an extra day off after the application of each of the jackets and except in the all-inclusive jackets of Butte, of Steindler and others, which sometimes require hospitalization.

3. With the removal of the corrective jackets and with the assumption of the retentive brace or corset, the intensive *exercises* begin. The brace is removed for exercises only. The patient's back is exposed, to observe every detail of the movement. The starting positions and the exercises follow (see p. 214). School children and working girls may continue their duties through the exercise period of the treatment; but a half hour should be allowed daily for the home exercises.

CORRECTIVE EXERCISES FOR LATERAL CURVATURE OF THE SPINE

In 1920, the late A. Mackenzie Forbes of Montreal, published a paper,[5] in which he described a "paradoxical position" in scoliosis, and amazed his colleagues by pointing out that attempted correction of the rotation deformity should be made in the direction exactly opposite to that in which it had been practiced. That is, in the right thoracic left lumbar type, correction was to be made by upper trunk-twisting to the right instead of to the left as heretofore. The writer satisfied himself that the twist of the "paradoxical" position lessened whilst the orthodox twist increased the rotation element, in which the rear right bulging of the ribs was the most obvious sign. Observations were made on many patients by *measuring* and recording, *semi-circumferences*, right and left, of the chest (at the level of the greatest thoracic deviation); in each case (Fig. 74b).

a. In the untwisted position of the upper trunk;
b. In the to-right twisted position; and
c. In the to-left twisted position.

The untwisted position (a) gave a right semi-circumference definitely greater than the left, as was to be expected; in the twist to the *right* (b), the discrepancy of the two sides was always distinctly *less*, and in the twist to the *left* (c), the discrepancy was

[5] Forbes, A. Mackenzie: Recent Studies in Scoliosis. Am. Med. Jl., July, 1920.

212   BODY POISE

always *greater*. The *X-ray tests* taken in the three positions named above were still more conclusive and afforded opportunity to note the relative twists of the vertebrae themselves (Figs. 72, 73, 74a). They also showed what happened relatively to the vertebrae in the lumbar deviation and the lumbar rotation. As

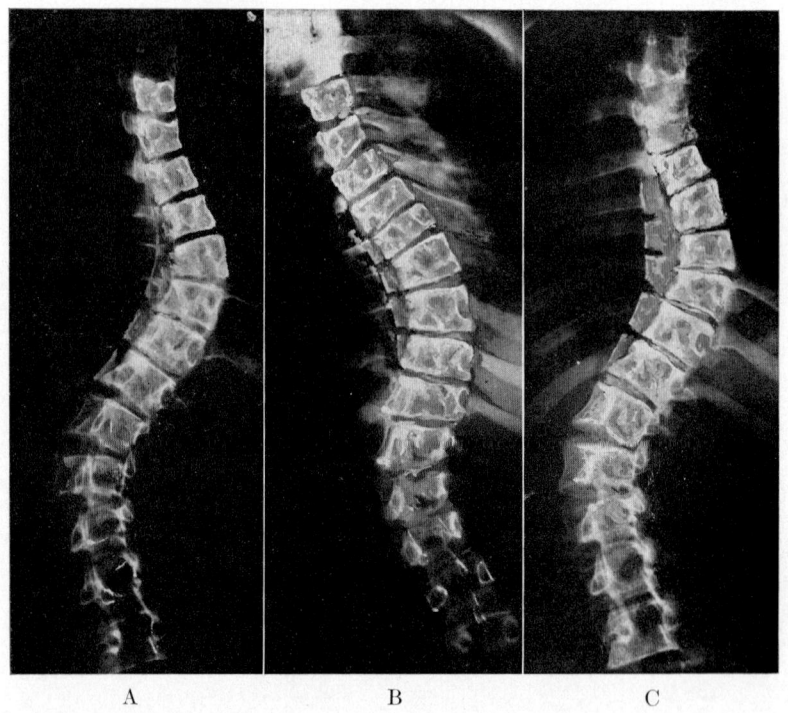

FIG. 72. Three X-rays of scoliotic spine, all taken with subject in prone-lying; that is, from rear forward.
  A. Without gymnastic twist of the upper trunk.
  B. With upper trunk-twist to the right.
  C. With upper trunk-twist to the left.

shown by the X-rays, in both regions, the deviations and rotations were lessened with the right upper trunk-twist and were increased with the left upper trunk-twist.

Forbes wished to apply this principle to corrective plaster of Paris jackets, but was obliged to state that he had not found the means of maintaining this correction in jackets. However, in

1922, Klein[6] published an effective means of using Forbes' principle with corrective plaster jackets.

The writer believed that the principle of the "paradoxical" position could be applied to corrective exercises and, in 1921[7] published and illustrated eleven exercises using this principle.

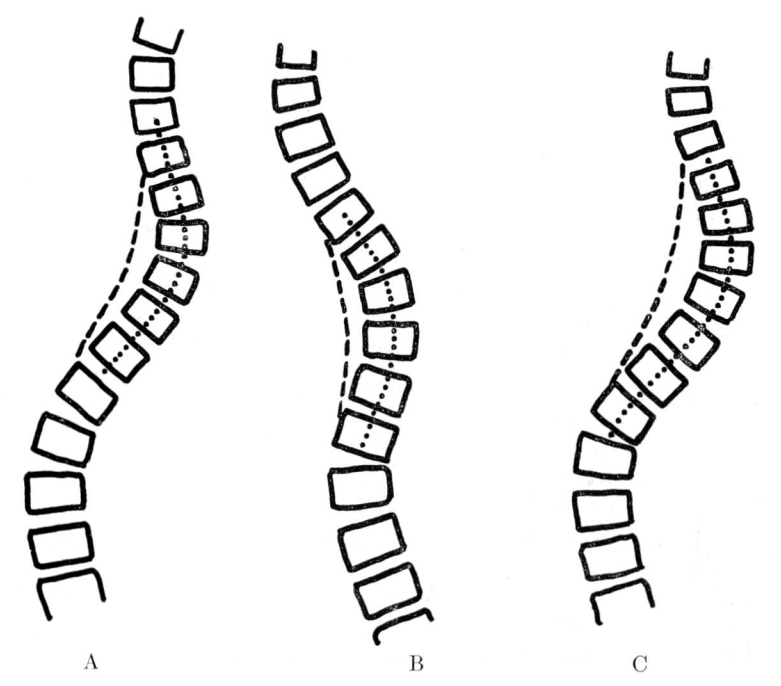

Fig. 73. Tracings of the spine only, of the three preceding X-ray photographs, to show (in broken lines) the positions of the successive rear-ward structures, and (in dotted lines) the positions of the centers of the succeeding vertebral bodies. Lateral deviation is greater in the bodies than in the rear-ward structures.

This group and others, since amplified from them, have brought *more rapid* and *more permanent* results than had been possible before.

The discussion of the rotation element has been given somewhat at length, in Part 1 (p. 70) of this book, because the entire inten-

[6] Klein, Armin: The Treatment of Structural Scoliosis at the Massachusetts General Hospital, Boston. Jl. Am. Med. Assn., Feb. 11, 1922.

[7] Truslow, W., Jl. of Orth. Surg., May, 1921.

sive corrective exercise treatment of scoliosis is based upon this conception of the deformity.

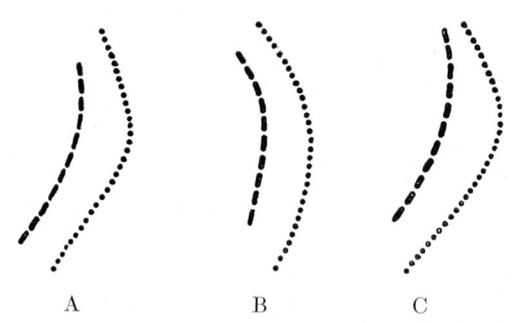

Fig. 74A. Repeating the three pairs of curves only. It is evident that, in position B, rotation (as well as lateral deviation) is less than in position A, and that, in position C, rotation is greater than in position A.

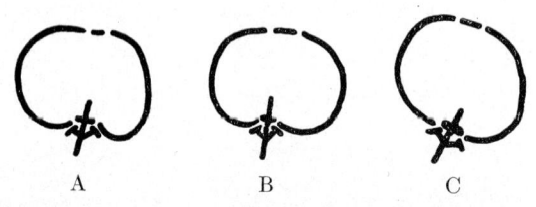

Fig. 74B. Tape measurements of right and left chest semicircumferences at level of greatest thoracic deviation.
 A. The deformity without upper trunk twist: R. 33.5; L. 26.5; discrepancy, 7.
 B. With upper trunk-twist to right: R. 31.7; L. 28.3; discrepancy, 3.4.
 C. With upper trunk-twist to left: R. 35.3; L. 24.7; discrepancy, 10.6.

EXERCISES FOR SCOLIOSIS

These are divided into A—Preliminary, B—Deformity Corrective and C—Posture Training.

A. Certain preliminary exercises are symmetrical and are used,
  a. To train the patient in the starting positions and in the simplest variations from them;
  b. To "limber up" stiffened muscles and ligaments of the trunk, the shoulder girdle and the hips; and
  c. To start the correction of the exaggerated antero-posterior curves.

Many of these are presented and their kinesiology discussed in Chapter 4.

A. PRELIMINARY EXERCISES FOR SCOLIOSIS

Kneeling

Ex. 1. With hands on hips, bend trunk forward until head touches floor (Fig. 53, p. 181).

Half kneeling—the left foot is placed forward on the floor.

Ex. 2. Bend trunk forward until trunk touches the knee.

Ex. 3. Bend upper trunk backward about thirty degrees (Fig. 54, p. 181).

On hands and knees

Ex. 4. Sway trunk forward to prone lying, then backward to resting on heels.

(Note: Do not change the position of the hands on the floor).

Prone lying

Ex. 5. "The seal"—with hands clasped low behind the back; raise head and shoulders (Fig. 56, p. 185).

Lying on back

Ex. 6. With knees drawn up (feet resting on the floor); bend both knees to the chest (Fig. 57, p. 186).

Ex. 7. With arms stretched upward beyond the head; arm flinging forward, raise trunk to sitting, to forward reach to the toes (Fig. 58, p. 187).

Half-prone lying at the end of a table

Ex. 8.—Alternate thigh-leg-foot raising to horizontal (knee straight) (Fig. 59, p. 188).

Ex. 9. Raising both thighs-legs-feet to horizontal (knees straight) (see Fig. 60, p. 189).

Ex. 10. With arms clasped low behind the back, raise head and shoulders.

(Note: Do not raise below the chest).

Sitting on a bench

Ex. 11. With feet apart and a dumbbell on floor between the feet; raise weight from floor to right shoulder, to high over head, to shoulder, to floor, to left shoulder, to high over head, to left shoulder, to floor (eight counts).

B. CORRECTIVE EXERCISES

The writer's intensive deformity-correcting exercises are progressively based on the preliminary group. They are distinctly asymmetrical. They aim definitely to correct or at least greatly improve the specific features of the deformity (see Fig. 66, p. 200) —the upper trunk-lean, the low shoulder, the compound spinal deviation, the exaggerated antero-posterior curves and especially the rotations. This is accomplished by the mechanism of the positions attained and by actively and progressively using and developing the muscles which must be depended upon to maintain such of these corrections as are attainable.

DEFORMITY CORRECTING (DE-ROTATION) EXERCISES

Kneeling

Ex. 1. With cane (or gymnasium wand) in hands; bend trunk forward to the left, reaching left side of cane far forward to the left on the floor, carrying right arm (bent at elbow) sideways upward, with upper trunk-twist to the right (Fig. 75).

In this exercise, the hips sway backward a trifle; the trunk bends forward and distinctly to the left; the left side of the cane rests perpendicularly on the floor forward to the left, with slightly bent left elbow; the trunk twists to the right, carrying the right arm bent with it, until the cane in the left hand is perpendicular, and the head and neck do not change this relative position to the trunk. This procedure shifts the upper trunk carriage to the left, corrects low left shoulder, lessens both lateral spinal curvatures and distinctly derotates the thoracic spine, with a relative evening of the asymmetrical rear rib walls (Fig. 76).

The mechanism of this and of succeeding de-rotation exercises is similar to that analyzed in Part I (see p. 70). The same four levers are used; the same initial powers—right Latissimus dorsi and other auxilliary muscles—are brought into play, and the same effect is basically produced; that is, the to-right twisted vertebrae

ROTARY LATERAL CURVATURE OF SPINE 217

Fig. 75. Ex. 1—Kneeling, right trunk-twist-bend

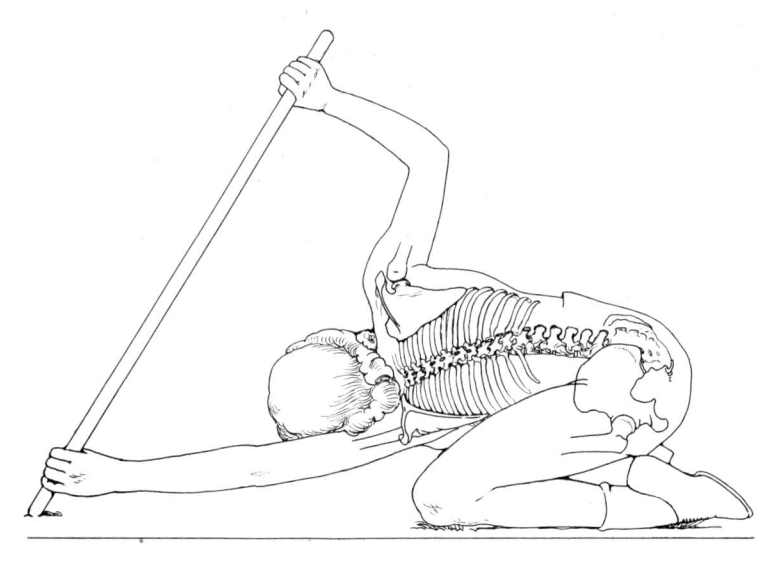

Fig. 76. Ex. 1—Analyzed to show changes in the spine—double lateral deviation changed to single, low left shoulder relatively levelled with the right and especially the thoracic rotations practically overcome. This study was made possible by the posturing of a flexibly mounted spine in the Anatomical Department of the New York Medical College. (Note: Similar spinal changes take place in all of the derotation exercises of this series.)

218   BODY POISE

are rotated toward the left and the asymmetrics of the chest wall are greatly lessened. It is to be noted that in mild cases the spinal rotations and chest wall deformities are often entirely overcome, for the moment at least, that even in severe cases these are much lessened and finally that with repetition of the derotation

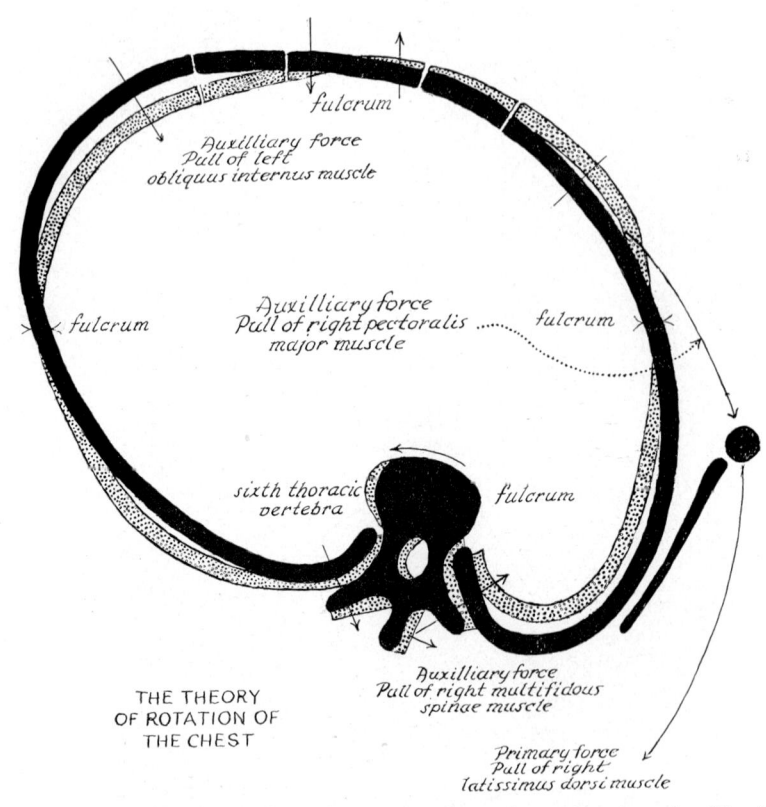

Fig. 77. The Theory of derotation of the ribs and vertebra (see also Fig. 17, p. 71) as applied to the chest deformity of scoliosis. (See text.)

exercises, ever increasing lessening of the deformities of rotation occur, with development of the muscles necessary to maintain such correction as may have been attained (Fig. 77).

On hands and knees

Ex. 2. Stretch right thigh-leg-foot backward and left arm forward (synchronous movement) (Fig. 78).

Fig. 78. Ex. 2—Kneeling, left arm stretch, right leg stretch

220   BODY POISE

FIG. 79. Ex. 4—Right stretch-trunk-twist, right leg-stretch, half-kneeling

Ex. 3. Place left foot forward on the floor and raise right arm sideways upward with upper trunk-twist to the right.

These exercises concentrate on the lessening of the lateral deviations—No. 1 lessens the lateral deviation of lumbar curve; No. 2 adds derotation to the thoracic curve.

Ex. 4. Stretch right thigh-leg-foot far backward, sway trunk backward (to sitting on the left heel), raise right arm sideways upward, twisting upper trunk to the right (synchronous movement) (Fig. 79).

To the derotating effect on the thoracic spine, this exercise adds lessening of the lumbar deviation to the left. To be properly taken, the extended right limb and right side of the pelvis should be forced backward as far as possible, causing a lessening of the lumbar deviation and necessitating concentric contraction of the Erector spinae muscle group of the opposite or left side of the low spine.

Prone-lying

Ex. 5. With left arm forward on the floor, head resting on left arm and with right arm stretched out sideways on the floor; raise the right arm sideways upward with upper trunk-twist to the right (Fig. 80).

(Note: Do not let hips move).

Later, with increasing dumbbell weight in right hand.

This again concentrates on the de-rotation element of the thoracic spine. It is essential that the trunk-twisting shall not include movement in the low spine or the raising of the right side of the pelvis.

Lying on the back

Ex. 6. With knees drawn up (feet resting on the floor); keeping knees parallel, bend knees toward chest, but twisting so that knees point to the right (feet to the left) (Fig. 81).

This movement lays stress on the lumbar curve to the left, with special reference to asymmetrical use of the abdominal muscles. The left side of the pelvis is drawn toward the left lower ribs, using the left Erector spinae muscles, the Obliquus

222   BODY POISE

Fig. 80. Ex. 5—Prone-lying, right-side-stretch, right trunk-twist

internus of the left side, the Obliquus externus of the right side and the Sacro-lumbalis of the left side.

Ex. 7. With arms over head on the floor; raise trunk to sitting, to left hand-touch to left toes and with right arm-raising sideways upward and upper trunk-twist to the right (synchronous movement) (Fig. 82).

This is a de-rotator of the thoracic spine, using especially the right Latissimus dorsi and left Deltoid muscles and with

FIG. 81. Ex. 6—Lying, double hip-knee-bend, with left twist

stress upon the development of the abdominal muscles concentrically in the first part of the exercise and of the back muscles eccentrically in the latter part. It also tends to correct faulty upper trunk-lean by the left hand-reach toward the left.

Half-prone-lying at the end of the table (Feet on the floor)
(Fig. 31, p. 132)

Ex. 8. With upper trunk carried to and resting near the left side of the table, left arm reaching far forward to grasp the table-top and with right arm stretched out sideways; raise right thigh-

FIG. 82. Ex. 7—Long sitting, right arm-stretch-trunk-twist, with left long reach

leg-foot to horizontal and raise right arm sideways upward with upper trunk-twist to the right (synchronous movement) (Fig. 83). Later, add increasing dumbbell weight in right hand.

The use of the table, in this exercise, affords better concentration on the de-rotating elements of the thoracic spine, and the flexed thighs lessen lordosis and hold the low spine and pelvis more fixed.

Fig. 83. Ex. 8—Half-prone lying, right arm-stretch-trunk-twist, with right leg raising.

Ex. 9. With forward stretched arms, hand-grasping both sides of table; raise parallel legs to horizontal, then a de-rotating twist to the low spine, by lifting the left side of the pelvis and carrying the parallel legs to the left, the left limb higher than the right. (Fig. 84).

The trunk rests on the table, with double hand grasping on the sides of the table, giving stability to the fixed upper half of the human frame. Although the Erector spinae muscle groups of both sides are used, concentrically from insertion to origin, that of the left side must exert greater action, both to raise the left side of the pelvis and to carry the pelvis and limbs to the left. Incidentally, the Glutei groups of both sides are used. This latter does not (for the moment) effect the de-rotation of the low spine, but adds to the strength of muscles necessary for good posture. (Note: This strenuous exercise should not be used until much previous training, in the table exercises, has been given).

Fig. 84. Ex. 9—Half-prone lying, double leg raising with hip twisting, left hip upward.

Left thigh support sitting on bench—"Spring Sitting"

Ex. 10. The left thigh is supported on the bench, the right thigh-leg-foot is stretched far backward, finger tips touching at each shoulder; bend the trunk forward to the left, reaching left arm forward (over left knee) to the floor, raising right arm sideways upward, with upper trunk-twist to the right (synchronous movement) (Fig. 85).

This exercise mechanically lessens all elements of the deformity of scoliosis. It carries the upper trunk-lean to the left and lifts left shoulder; the body-stretch tends to straighten both lateral deviations; the backward stretch of the right limb fixes the correction attained in the lumbar deviation, and finally the twist of the upper trunk, with right arm carriage sideways upward, de-rotates the spine. Although many muscles are used, the principal motive factor in de-rotation is the concentric action of the Latissimus dorsi muscle of the right side.

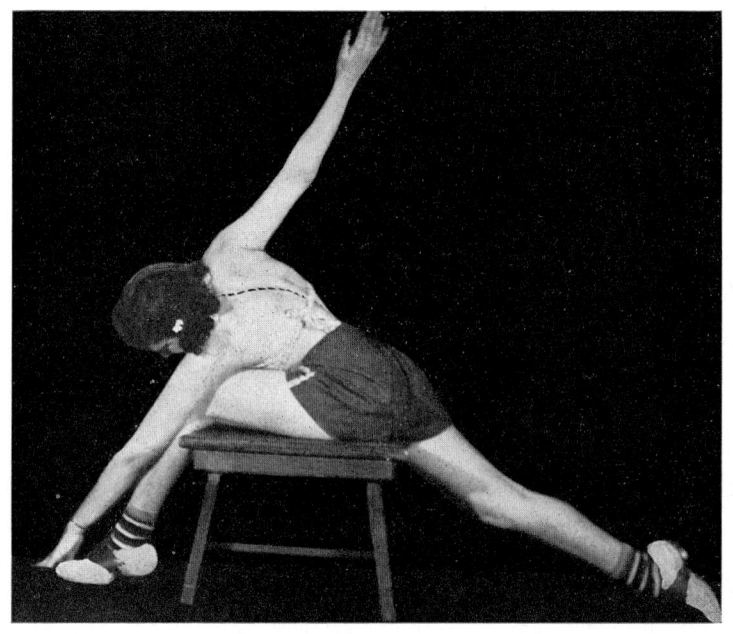

Fig. 85. Ex. 10—Left spring-sitting, right arm-stretch-trunk-twist, with left arm forward downward.

Ex. 11. Left hand-support "spring sitting." The left hand rests on a table (far forward), the remainder of the body is in spring sitting with right arm hanging near the floor; raise the right arm sideways upward with upper trunk-twist to the right.

This exercise concentrates upon the de-rotation element of the deformity, while mechanically improving all other elements by the use of the starting position.

When the above list is thoroughly understood and much correction has been attained by its use, an intelligent patient may be shown a single *Key-note corrective exercise* with which to carry-on the corrections attained and to add thereto.

Ex. 12. Key-note rotation exercise.

Starting Position—Standing on right foot, left foot placed on a bench, elbows bent with fingers touching shoulders.

The movement—Bend trunk sideways forward, with upper trunk-twist to the right, and at the same time straighten and carry the left arm forward and downward *over* the left knee and with the straight right arm carried sideways upward (Fig. 86).

(Note: The patient is instructed to endeavor to get the left arm-pit over the left knee—neither to the right nor to the left of the left knee. With much deformity this cannot be attained at first, because the trunk deformity lessens the spine length to such an extent that it cannot reach the thigh-length. The patient may be told, however, that if and when the left arm-pit completely covers the left patella, it indicates that the "kinks" in the spine have been much straightened—that a corrected trunk-length has practically equalled the thigh length for that individual).

(For study of the kinesiology of this exercise, see text following the next one.)

Ex. 13. Modified Key-note exercise—To meet a reasonable criticism that the foregoing exercise may place undue strain upon the supporting right foot, the exercise may be modified by substituting half (or left-thigh) support-sitting at the end of a table, and carrying the movement through as outlined above (Fig. 87).

In both of these Key-note exercises, all elements of the deformity are improved, some momentarily over-corrected. The faulty trunk-lean and low left shoulder are markedly over-corrected; with the left arm, as a lever, over the left knee as a fulcrum, the two lateral deviations of the spine are materially lessened, the low-carried right limb (especially in the Modified Key-note exercise) adding to the improvement in the lumbar deviation, and finally the trunk-twist and right arm carriage sideways upward, furnishes the de-rotating element. The right Latissimus dorsi and component muscles constitute the active force in de-rotation of the thoracic spine and lessening of chest wall distortions.

Note: This Modified Key-note exercise, taken from the table-support position, affords the physician or physical trainer an

FIG. 86. Ex. 12—Half-standing, left bend right arm-stretch-trunk-twist, with left arm forward downward. (Writer's Key-note exercise.)

excellent vantage point from which he may add his own manipulations to assist in the elongation and straightening of the spine.

FIG. 87. Ex. 13—Left high-sitting, left bend right arm-stretch-trunk-twist, with left arm forward downward. (Writer's Modified Key-note exercise.)

To do this, the operator stands facing the table-supported knee. He grasps the patient's left wrist and pulls the left arm (and trunk) forward over the patella and then gently downward, using the patient's left arm as a lever to carry the arm downward in front of the patella. When as much of this as seems best is accomplished, still holding the left arm-pull, the operator grasps the patient's right elbow and assists the right upper trunk-twist. This outside aid should augment, not replace, the active use of Latissimus dorsi and other muscles by the patient (assistive exercise).

C. POSTURE TRAINING IN SCOLIOSIS

Fairly early in the exercise treatment of scoliosis, posture training should be started. An experienced corrective gymnastic instructor will have his own methods, but the following exercises emphasize the elements of deformity to be overcome, and have been found useful. The patient stands before a wall mirror. He is shown and trained to take as good an attitude as may be. From this stance, the following changes in the position of the trunk, of the arms and of the legs are made:

1. With hands on hips (hips firm), carry the head, shoulders and trunk to the left—emphasize "carry left shoulder upward and to the left."
2. With left hand behind neck, right hand on hip, carry head, shoulders and trunk to the left and upward (Fig. 88)—Bernard Roth's Key-note exercise.
3. With left arm raised over head, right arm sideways outward, carry head, shoulders and trunk to the left and upward.

> (In these three exercises, the hips (pelvis) are not carried to the left. Indeed shifting the levelled pelvis slightly to the right will maintain the general body balance and will emphasize the importance of the *lateral trunk pull away from the hips*).

With left hand behind neck, right hand on hip and with upper trunk carried to the left (see Ex. 1 of this series), two more exercises may be given by changes in the legs:

4. Raise heels, bend knees, extend knees, sink heels (toe-knee-bend standing) (Fig. 64, p. 195).

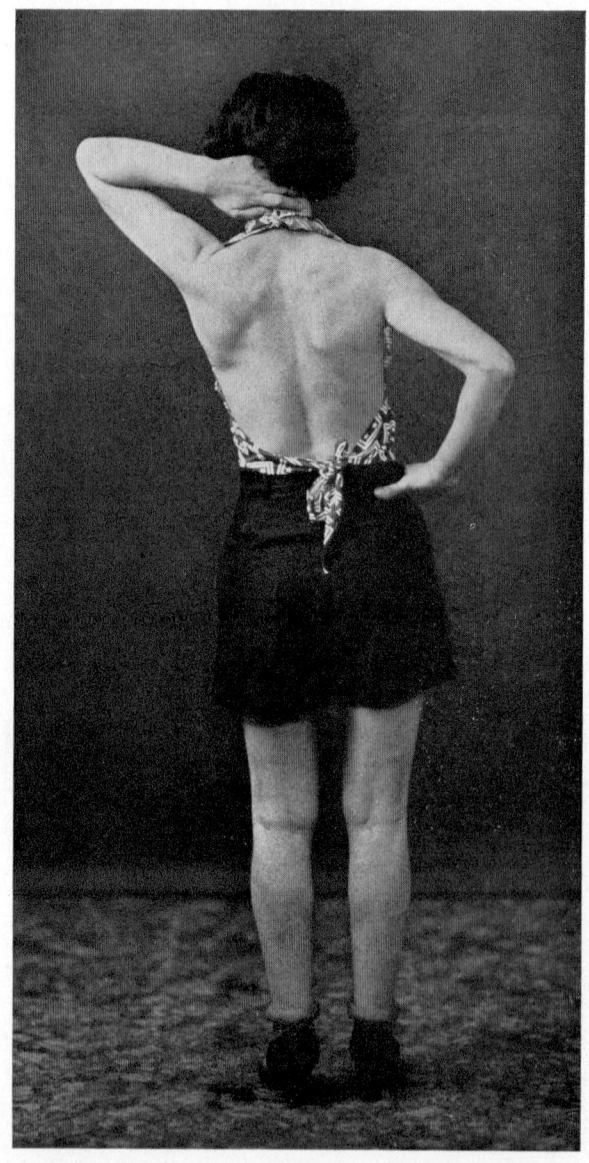

Fig. 88. Posture training. Left neck right wing standing, with upper trunk pull to the left. (Bernard Roth's Key-note exercise.)

5. Raise left heel.

(In both of these, especially in number 5, stress should be laid on the continued pull of the trunk to the left, the raised left shoulder and on holding the abdomen well in, while continuing easy chest and diaphragm breathing).

After there has been considerable training in these posture exercises, additional corrected poise may be added:

6. Upper trunk in the postures attained by exercises 1, 2 or 3 (above) and with a soft but moderately heavy sandbag on top of the evelled head; walking away from and return to the wall mirror to check up:

    a. In ordinary march step;
    b. In high-knee-bend march step.

One must remember that posture training is mental as well as physical. The efficacy of it depends largely upon the trainer, upon his ability to "get it across" to the trainee and upon the latter's ability to "take it in."

CLASSWORK CORRECTIVE GYMNASTICS

It is obvious that the details of all of the exercises here presented, whether the intensely corrective or the postural, are so important that the best results will be attained by work with the individual. But where a gymnasium or a large room is available and where persons are well grounded in individual work, carry-on may be given in groups made up of not more than six persons. The writer has had classes in scoliosis to as many as twenty persons, but he was aided by trained assistants who went about helping him in the correction of individual faults.[8] The class method has the advantage of added interest to the patients and of economy of time and of cost to the patients. It has been found that groupings are best made by age and sex rather than by physical deformities.

D. OPERATIONS FOR SCOLIOSIS

Orthopedic surgeons today, in increasing numbers, are recommending *spinal fixation*, in the best position attainable, by

---

[8] Truslow, W.: Medical News, N. Y., Apr. 26, 1899.

bone grafting or bone fusion operation on the spine itself, for those *paralytic* cases in which the intrinsic muscles of the spine are so markedly involved as to expect little or no "come back" by exercise development. Some advise this for non-paralytic structural scoliosis as well. It is a severe operation and should not be undertaken without weighing carefully the many factors involved. And yet the condition of the greatly paralyzed patient, whose deformity often increases under the most careful "external bracing", is so appalling, and the results by operation have been so brilliant, in the hands of experienced surgeons, as to justify it. The procedure includes a preliminary two to four weeks in bed with increasing longitudinal stretching, by pulley weights applied to the head, to feet and pelvis and with lateral tractions at the levels of greatest lateral deformity; or a few weeks use of the rapidly correcting hinged segmental plaster jacket; and the operation, which is often done in two stages—that is, a segment of perhaps six to eight vertebrae fused or bone grafted, waiting three to six months before fusing of another six to eight vertebrae, with retentive plaster jackets for about one year. The pros and cons of this method of attacking severe paralytic scoliosis—and the writer does not endorse it in the non-paralytic curvature of the spine—are scarcely within the scope of this book, but they are important and must be discussed, for the individual case, with one most experienced. But one may note that after a spine is fused, intensive corrective exercises are not given. Indeed they are decidedly contra-indicated.

ESTIMATE OF METHODS OF TREATMENT

No real improvement in structural scoliosis can be made unless a well balanced use of rapid correction with plaster jackets and retention support and corrective and postural exercises are carried through. In the mild functional cases, with no anatomical variations, real cure may be obtained by the use of exercises, often with a short time use of a retention brace or corset; in the structural cases improvement only, often marked and—what is important—estoppage of deformity progession is all that has, as yet, been attained.[9]

[9] Report of the Scoliosis Committee of the American Orthopedic Association.

A. A corrective jacket lessens the deformity more rapidly than does brace-wearing or exercises. It affects rotation least of all of the elements of deformity. It has distinct time limitation because of skin pressure intolerance and because of the patient's attitude toward it. It should be resumed after a shorter interval of bracing and exercises, in paralytic spine patients.

B. The retentive brace or corset alone should retard deformity formation. It will bring about no correction of it, and unless constantly cared for will allow increase in deformity. It is usually inadequate in paralytic spine.

C. Exercises alone will not be sufficient to prevent an increase in a deformity in which the ratio of lateral deviation deformity is greater than four percent (see p. 205). They can be used only with very gradual progression in the paralytic. When reinforced by an efficient retention brace and intermitted with an occasional short return to the corrective jacket, it is the best means available for insuring a stoppage of deformity progress, for obtaining a large amount of deformity lessening and, by its general hygienic as well as local effect, for a reasonable assurance of non-return of deformity. The intensive exercises here used are very effective especially in lessening the rotation elements of deformity.

As to *time necessary*, one would say that a structural scoliosis presenting five percent scoliosis or less would require about one year of active treatment—plaster corrective jackets for three months, nine months of retentive brace or corset with intensive supervised exercise—; and that in the second year the patient could wear a simple reinforced corset and do home exercises daily, with occasional professional supervision. A ratio of deviation of five to ten percent would require three months of corrective plaster jackets; six months of retentive brace or corset and intensive exercises; a second three months of corrective plaster jackets, and a second year of bracing with supervised exercises. Greater amounts of deformity should require longer time. The paralytic, if treated non-operatively, must have a larger proportion of the time given to the corrective plaster jackets and must be carried on for several years.

D—Internal Fixation, by operative spinal bone grafting or spinal splinting, certainly has its place in the treatment of scoliosis,

The indications are inability to prevent increase in deformity by non-operative means and danger to the patient's health—because of mechanical restriction to the lung or heart functions—and finally weakening of the patient's morale from previously long and ineffectual treatment. Because an operatively fixed spine is subject to little or no post-operative improvement in the deformity, as much *pre-operative redressment of the deformity* as is possible *is* (and should be) *attained*. While recognizing the possibilities of the controlled segmented plaster jacket, as a deformity corrector, but noting first that such jackets do not sufficiently overcome the rotation element and secondly that the writer's derotation exercises are potent factors in lessening rotation, a few weeks of intensive de-rotation exercises should be a part of the pre-operative preparation of the patient for the internal fixation operation. If then a controlled segmented plaster jacket is also used, its application will more fully correct lateral deviations, faulty body-lean and low shoulder, and the surgeon may feel that he has done all that is possible to insure a spine and trunk that will be as much corrected as can be.

SUMMARY

1. Successful treatment of structural scoliosis must depend upon a clear understanding of the elements of deformity, and the lessening if not complete elimination of them all.

2. Roentgenological preliminary examination of the spine in three body positions is an aid in the study of the deformity. But repeated regular measurements and numerical recording of the elements of deformity are important as guides to continuance of treatment and as indicating elements most needing correction. A simplified method for doing this is explained.

3. A balanced use of corrective plaster of Paris jackets, of retentive brace or corset and of intensive exercises is essential to satisfactory results.

4. The position of the patient when the plaster of Paris jacket is applied is responsible for improving body posture and shoulder carriage; the successive paddings are the means used for correcting spinal deviations and rotations and for reshaping of the bony chest.

5. Essentials of a retention brace are:
   a. Ability to hold corrections attained;
   b. Application by the patient with reasonable accuracy;
   c. Extensibility and lateral compressibility to meet normal growth and progressive deformity decrease;
   d. Mechanical further self-correction by a brace seems possible but not yet attained.

6. Gymnastic corrective exercises must be progressive, intensive and at first with a minimum of weight bearing. They must aim to lessen all of the elements of deformity, especially that of rotation. Starting positions other than the standing facilitate this. The author's de-rotation exercises are herein given, even to the exclusion of other valuable ones, in the belief that they more nearly meet the requirements set forth than do any others and bring quicker and more lasting results.

7. Lessened deformity attained by corrective jackets must be maintained by retention braces or corsets while muscle exercise is developing natural muscle support. Artificial support may gradually give way to natural support. The paralytic scoliotic must receive a larger proportion of artificial support than is used for those not paralyzed in the trunk muscles.

8. Internal fixation, by operative bone-fusion or grafting, may be necessary in severe paralytic cases, but must not be undertaken except after thoughtful consultation with an experienced operator. Intensive corrective exercises are definitely contra-indicated after such a fixation operation, if it has been successful in accomplishing firm fixation. Therefore the greatest pre-operative correction of the deformity as possible should precede operation. The author's de-rotation exercises have its place in this.

In this book, Body Poise has been studied first from its Anatomy and Kinesiology. This has included the writer's anatomical explanation of Mackenzie Forbes' "paradoxical position" in de-rotation of the spine. Then three abnormal conditions, affecting Body Poise have been presented by definition, by measurement of deformity and by treatment. As derangements of foot balance are outstandingly important in entire trunk balance, this subject has been given first; the simpler functional derangements of the spine and trunk have followed in the discussion and finally, the severer structural deformities of Scoliosis have been studied. In all of this, the entire treatment has been outlined, some of it in detail; but stress has been placed upon Corrective Exercises, because it is a common and often neglected part of the treatment of all of the conditions presented. In the gymnastic treatment of Scoliosis, the writer's de-rotation exercises were enlarged upon as the quickest and most effective mechanical means of lessening the deformity.

Part III
BODY POISE IN GAMES, SPORTS AND ATHLETICS

INTRODUCTION

Games, Sports and Athletics are indulged in as pure pastime, as means of body building and, especially in Anglo-Saxon countries, as inter-group contests. Although the three terms are often used interchangeably, there is some distinction in their meaning. In Webster's Dictionary, games are defined as single contests lasting until a definite limit is reached; sports, as pastimes which may be carried on by individuals or in groups, and athletics are the games and sports of athletes, the word specifically involving the art of training athletes. Although the United States is classed as an Anglo-Saxon country, many of our best athletes are of stock other than that racial group. The spirit of our original ancestors, however, has been carried on in other racial strains that are now integral parts of our population. So we see our people today as members of a sport-loving Nation, and are rather proud of it. This is also the British view-point. In the countries of Continental Europe, stated gymnastics have the edge on sports and games, although athletics are making much headway.

What is the effect on the health of our people—our boys and our girls, our men and our women—of this emphasis on sports and games as our principal means of physical exercise? On the whole, good, but there are some adverse tendencies which should be regulated.

In Part I of this book, the ideal of good body posture has been analyzed, anatomically and kinesiologically, and, in Part II, three definite departures from good posture have been studied and their remedies enumerated. Part III will carry on the principles of Part I and Part II as they relate to a number of popular games and sports. In doing this, one is confronted with the well-nigh universal and laudable ambition to excel, by special body training.

Sometimes this spirit of emulation leads to ill-proportioned body development. Fortunately, however, the idea of well-proportioned and well-conditioned physiques is growing among our better informed physical trainers and coaches and indeed among the athletes themselves. It is the basis of the training of registrants in the Army, in the Navy and in the Marine Corps, where weeks of general, symmetrical physical exercise precedes special training in specialized military activities and needs. It should be the basis of training of contestants in sports and games.

To know what is good and what is harmful in our popular sports, we should study them thoroughly. That is the theme of this section. Each sport listed will be treated with a brief historical sketch, an analysis of the kinesiology of some of its leading sport positions—as, for instance, the pitcher, in baseball—mal-proportioned and asymmetric body tendencies will be pointed out, and finally some suggestions as how to remedy or counteract the harmful results of these tendencies will be made.

Although other adverse physical conditions, as for instance heart strain, may be involved, the relations of the sports and games to body poise will be the criterion of these analyses. In general, the faulty postural tendencies in specialized sports and games may be classified as (a) horizontally disproportioned muscular and structural development and (b) asymmetrical or lateral disproportion. The latter involves excessive use of right shoulder, arm and hand—or of the left in so-called left-handed persons. Examples of horizontal disproportion are boxing, canoeing, running, skating, cycling. Examples of asymmetrical or lateral disproportion are baseball, football, golf, tennis, shot-put, javelin throwing, and they are due to the preponderance of uni-dexterity —right handedness in the majority, left handedness in the few— of the contestants.

The origin of right-handedness and of left-handedness is obscure. Why are the majority of persons, estimated as 92 to 96 percent, right-handed and the remaining 4 to 8 percent left-handed? There are two theories—the acquired and the congenital—and a third or combination of the two. Those who believe in the acquired theory, put it down to social training and imitation, stating that the child is early taught to use the right hand in eating and in other daily functions and that imitation and later an innate abhorrence of a social anachronism fixes the habit. The advocates of the congenital theory base their findings on the structure of the human body—a larger left side of the brain (with crossing of the nerve fibres in the medulla oblongata to account for right-handedness), asymmetrical distribution of body viscera with lateral displacement of the center of gravity of the body, and finally supposed inequality of blood supply to the two halves of the brain. The intermediate theorists find the origin of right-handedness in primitive warfare, the club or spear wielded by the right hand so that the left-arm-carried shield may protect the heart. These theories are presented without comment. For our purposes, we find the vast majority of the human race is right-handed, and will consider the trait as a prime factor in the tendency to lateral disproportion in muscular and structural development.

Sports and games are herein *classified* as:

I. Those which present little or no bad tendencies and indeed aid good posture. They include walking, swimming and canoeing (if the paddle is used equally on alternate sides);

II. Those which tend mildly to faulty posture which can be counteracted. Among these are rowing (with sliding seat), boxing, basketabll, running;

III. Those which present marked tendencies to horizontal disproportion or to marked lateral asymmetries in development, which should not necessarily bar their use but which should be undertaken with counteracting physical work. Of these are enumerated baseball, football, golf, tennis, hurdling, cycling, shot-put, javelin throwing, archery.

In this section, stress will be placed upon preliminary and follow-through general calisthenics, or "setting-up" drills, as a principal means of counteracting ill proportioned results. In some cases other sports, contra-lateral indulgence in the same sport and, occasionally, corrective gymnastics will be suggested to attain the same end.

Chapter 6

I. SPORTS AND CONTESTS PRESENTING LITTLE OR NO TENDENCIES TO POOR POSTURE

WALKING

Walking as a function of man, of course, traces far back to pre-historic times, to a remote period, when man received the name (given him in modern times) of *homo erectus;* and through the ages since those primitive days to the present, walking has been the most universal method of locomotion of *homo sapiens.* Indeed, wise is the man or woman who, in these days of mechanical transportation, does not neglect this health-giving function!

In historic times, walking has been practiced by most people, but one of the earliest records was that of Paul of Tarsus, in the first Century of the Christian era, who "sent his company by ship" as he "minded himself to go afoot". Rousseau, Thomas Carlyle, John Bryce, Wordsworth, Emerson and Thoreau have written concerning the joys of walking and of mountain climbing. Leary and Weston, in the second half of the Nineteenth Century, were famous as long distance walkers. Walking, as a form of competitive racing, does not appear to have been popular until the latter part of the Nineteenth Century. In the English championship games of 1866, J. G. Chambers covered the seven-mile race in 59 minutes, 32 seconds. In the American championship games of 1876, there were one mile, three mile and seven mile walking events. In the Olympic games walking was sometimes included and as often dropped. The one mile walking race has continued most consistently.

THE KINESIOLOGY OF ORDINARY WALKING

Walking may be defined as the horizontal projection of the human frame in space, from alternate foot placements on the ground, the bone, joint and muscle mechanism, from the feet upward, being the active factors. In ordinary walking, the working mechanism is confined to the hip joints, thighs, knee joints, leg bones, ankle joints and feet, with slight compensatory swinging of the pendant arms. As effort increases, some lateral twist of the pelvis and increased arm swinging occurs; until, in race walking, the entire torso, hips and arms add much motion to increase the effectiveness of the exercise.

Walking is distinguished from running by certain relationships of the feet to the ground. In walking, one or the other foot is always in contact with the ground and both feet may momentarily be in contact for short periods. On the other hand, in running, both feet are never simultaneously in ground contacts and there are successive short periods in which the body is entirely free from the ground, as alternate forefoot contacts initiate springings from the ground. Six illustrated phases, in walking are analyzed:

Phase 1 (Fig. 89). The right foot is firmly on the ground, the right leg is backward, the left leg is forward with the foot off the ground. The right Glutei muscles are in concentric contraction; right Quadriceps muscles in concentric contraction; right calf muscles in eccentric contraction; left Ilio-psoas muscles in concentric contraction; left Quadriceps in concentric contraction; and right foot flexors (dorsi-flexors) in tonic contraction.

Phase 2. The right foot is leaving the ground with the toes still in contact, the right leg is backward, the left foot is forward, the heel being in contact with the ground. This is an unstable position, but is maintained by the momentum of the body. The right Glutei muscles are in concentric contraction; the right hamstring muscles are in concentric contraction; the right calf muscles are in tonic contraction; the left Ilio-psoas muscles are in concentric contraction; the left Quadriceps are in concentric contraction; and the left foot dorsi-flexors are in tonic contraction.

Phase 3. The right toes still touch the ground, the right knee is bent, the right hip joint is in lessened extension, the left foot is firmly on the ground, the left knee is straightening and the left hip joint has lessened flexion. The major effort is being transferred to the left leg. The left Glutei muscles are in eccentric contraction; the left Quadriceps muscles are in eccentric contraction; and the left hamstring muscles are in concentric contraction (the left hams giving up as the left front-thigh muscles take on the effort); the left foot dorsi-flexors (working from insertion to origin) are in concentric contraction, helping to carry the left leg and thus the entire trunk forward; and the right

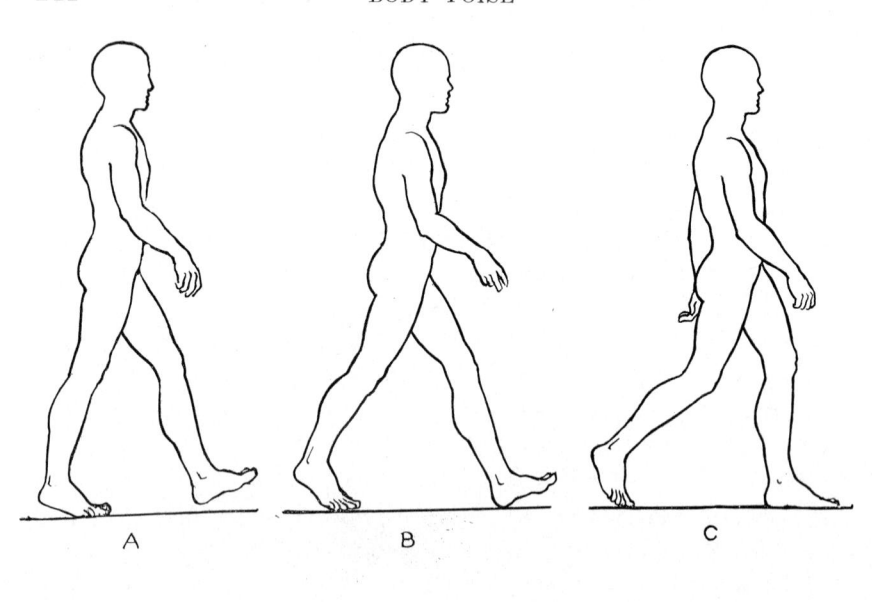

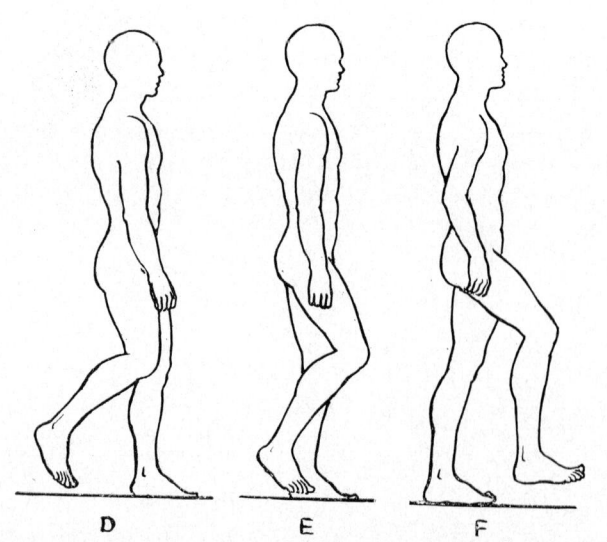

Fig. 89. Mechanism of walking—six phases (redrawn, Encyclopedia Britannica).

Ilio-psoas, right Quadriceps and right calf muscles are in contraction, preparing to lift and carry the right limb forward.

Phase 4. The left foot is still firmly on the ground and the left knee is fixed in full extension, while the right hip joint is in neutral position, the right knee joint is flexed and the right foot is carried off the ground. The left Glutei and the left hamstring muscles and the left dorsi-flexors of the foot are in concentric contraction, to fix the left limb and prepare for the carrying of the trunk forward. The left Quadriceps are in contraction to maintain the fully extended left knee; the right Ilio-psoas muscles are in concentric contraction, to carry the right thigh forward; and right Quadriceps are in eccentric contraction, to begin the extension of the right knee; and the right dorsi-flexors are in tonic contraction to prevent right "foot drop".

Phase 5. There is but little change in the supporting left limb. The right hip is flexed; the right knee is still flexed and the right foot is maintained in neutral position. The muscle status of the left limb is similar to that of Phase 4. The right Ilio-psoas is in concentric contraction; the right hamstrings are still in concentric contraction, holding the knee bend; and the right dorsi-flexors are used to hold the foot in neutral position.

Phase 6. The left hip is in moderate extension, the left knee is fully extended and the left ankle is in moderate dorsi-flexion, from which points the trunk is carried forward; the right pendant limb is carried forward with flexion at the hip and at the knee and with neutral position of the ankle. The left buttocks' muscles are in concentric contraction; the left Quadriceps in tonic contraction; and the left dorsi-flexors of the foot are in concentric contraction; the right Ilio-psoas muscles are in concentric contraction; the right Quadriceps are in eccentric contraction and the right dorsi-flexors are in tonic contraction.

From this point, the cycle of walking begins again.

Ordinary walking should present *no hazard to good body poise*. Indeed it is to be considered as an aid, if thought is given to erect body carriage while walking. The only suggestion to be made is to advise gradual progression as to distance, and, to the distance walker, five minutes complete rest upon the back, for every half hour of walking.

246   BODY POISE

THE KINESIOLOGY OF RACE WALKING

Race walking differs from ordinary walking to some extent, especially in the greater length of the stride and in the activity of the trunk and arms. But one phase will be described—that of the longest stride distance, with the left limb far forward and the right limb far backward—with some comment on the relationship of this phase to succeeding phases (Fig. 90).

Fig. 90. Mechanism of race walking—phase of greatest stride distance (redrawn, Encyclopedia Britannica).

The heel of the left foot and the toes of the right foot are simultaneously though unstably in contact with the ground, both knees are fully extended, the left thigh is flexed and outward rotated, the right thigh is extended and inward rotated, the pelvis is twisted forward on the left side, the trunk is twisted forward on the right side, the left half-elbow-bent arm is carried backward (extended) and the right half-elbow-bent arm is carried forward (flexed at shoulder joint). The left Quadriceps extensor muscle is in concentric contraction to hold the knee joint straight; the dorsi-flexors of the left foot are in sufficient contraction to hold the toes upward; the left Ilio-psoas muscle and the left Quadriceps (Rectus origin) are in contraction to flex and out-rotate the left thigh. The right calf muscles are contracted, to raise the right heel; the right Quadriceps extensor muscle holds the right knee extended; the right Glutei and Hamstring muscles (acting from insertion to origin) have extended the right hip and carried the pelvis forward, and the right hip in-rotator muscles have twisted the pelvis forward on the left side.

It is to be noted that, to this point in the description of the kinesiology of the long-stride phase of race walking, the right Glutei muscles and the right hip in-rotators are the most effective muscles used to attain forward progression of the lower half of the walker.

In the description of ordinary walking, the motor mechanism above the pelvis was not discussed, because the body parts above the pelvis are, in ordinary walking, carried as an almost static load on the pelvis, the mechanism of good posture only (with perhaps some arm swinging) being all that was required for these upper segments. But in race walking, the end aimed at is the most expeditious attainment of forward progression of the entire body. To do this, in the cycle of race walking, torso-twisting and arm movements are also used. They are employed, as will be seen, both to aid forward motion of the upper segments and as momentary anchorages, at each phase, to further the effectiveness of the next phase.

So, to return to the torso, arm and head mechanisms of the longest stride-distance phase, in race walking, the torso is twisted forward to the right, using the left Internal oblique and the right External oblique muscles of the abdomen, the many long and short rotators of spine—some on the right side and some on the left— and the powerful Latissimus dorsi muscle of the left side; the left scapula is adducted by the left Trapezius and Rhomboid muscles; the left upper arm (humerus) is extended by the left Latissimus dorsi and rear third of Deltoid muscles; the left forearm is semi-flexed by the Biceps and other elbow flexor muscles; the right scapula is abducted by the right Pectoralis major muscle; the right upper arm (humerus) is flexed by Pectoralis major and anterior third of Deltoid muscles, and the forearm is semi-flexed by the Biceps and other elbow flexor muscles. The Head and neck maintain the position of the forward progression of the contestant throughout the entire cycle of race walking. This muscle-joint mechanism of structures above the pelvis has materially aided in the advancement of these body segments, particularly advancing the right side of the upper torso in the phase analyzed.

Without analyzing the details of the next succeeding movement, it is sufficient to state that, the positions thus attained throughout the entire human frame become the momentarily fixed points for the kinesiological mechanism of the next movements, which for the most part use opposite muscle mechanisms; and that, in each succeeding phase, the same principles of sequence of muscle action and joint mechanism prevails.

As in ordinary walking, race walking presents *no hazard to good posture*. It is practically a universal body developer and, as the unilateral and twisting efforts are completely neutralized by identical contra lateral muscle work, there is no tendency to body asymmetry.

SWIMMING

Swimming is the art of propelling the body through water without the use of extraneous aids. Methods include the breast stroke, the crawl, the side stroke, the back crawl and swimming under water. This analysis will be made of the breast stroke, with comments on other forms.

Swimming certainly antedates historic man. Most persons learn to swim. There are records of automatic swimming without special training—of boys, with no previous experience, being thrown into deep water and by arm and leg strokes (at least simulating those of good swimming) arriving without help at the safety of solid ground or of a boat. But for most of us, it is taught. Swimming has always been used as a pleasurable exercise or as a means of crossing a stream, a lake or a bay. Great Britain for many years held the records for contest or speed swimming; but America, Australia and Continental Europe have in later years vied with Britain's supremacy. Swimming for long distances has had its devotees. The English Channel from Dover to Calais and the Hudson River from Albany to New York have been favorite courses.

KINESIOLOGY OF THE BREAST STROKE (FIG. 91)

The start and the cycle, in four counts, will be analyzed. The *start* is the dive in or the push off, with a momentary glide in the water, with straight arms in front and straight legs behind. The *counts* are: (1) arms carried sideways to shoulder level, no leg movement; (2) elbows flexed to sides of chest and bend and spread of knees; (3) arms shot forward, knees straightened and feet far spread; (4) legs snapping together, arms held in forward position.

The Start

The swimmer either dives horizontally from the end of the pool or pushes, with his legs, to propel the body as far forward on the surface of the water as he can. His arms are fully extended at the sides of his head, with palms of the hand and his extended fingers together, and his legs are extended backward and feet plantar flexed, to make as efficient a "stream-lined" glide forward as he can. The head may or may not be out of water.

The leg motion of the plunge has but little anatomical significance, except to point out that vigor and precision in the extension of both hips, knees and ankles, and immediate attainment of the position of the torpedo-like glide has much to do with "getting a good start". On attaining the prone position, the buoyancy of the water minimizes the necessity of activity of the great trunk muscles. But both shoulder-blades are rotated upward, requiring use of the Trapezii and the Serratus magnus of both sides; the arms are fully abducted, using both Deltoid and Pectoral muscles; the extension of elbows is maintained by the Triceps, the semipronation of the forearms requires activity of Pronator radii teres and Pronator quadratus of both sides. The Gastrocnemii hold both feet in plantar extension. A short-timed glide, to take advantage of the momentum of the spring-off start, precedes the muscular motions of swimming.

Count[1] 1. *The extended arms are carried downward to shoulder level. No leg action.*

---

[1] Of course no counting is done, but this naming of synchronized movements aids description.

250   BODY POISE

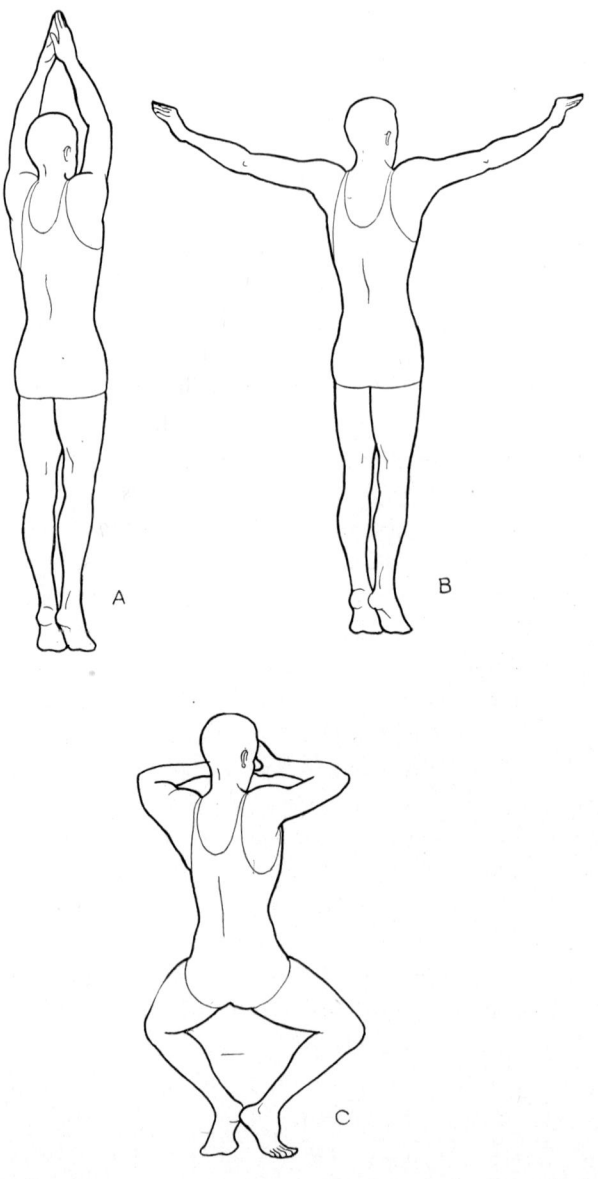

Fig. 91. Mechanism of swimming, breast stroke—three phases (redrawn, Encyclopedia Britannica).

There are no changes in the mechanism of the trunk or of the legs and feet. The shoulder-blades turn from extreme to lessened upward rotation (see p. 112); the upper arms are brought from extreme abduction to half abduction and to inward rotation; the forearms from semi-pronation to full pronation and the hands (with fingers closed, to get as good paddle effect as may be) face the pushed-back water.

Latissimus dorsi and lower half of Pectoralis major muscles de-rotate the scapulae and adduct and inrotate the upper arm; the Triceps hold the elbows extended; the Pronator radii teres and Pronator Quadratus complete the pronation of the forearms; the wrist flexors keep the hands from dorsi-flexion, and the adductors of the fingers and the thumbs maintain the hands as "paddles"[2] and the extended arms as "sweeps".

Count 2. *The elbows are flexed at the sides of the chest and the knees are flexed and widely separated. Beginning of the "frog kick".*

There is no pause in the movements of the arms. The shoulder-blades are adducted and have no upward rotation; the upper arms are adducted; the elbows are sharply flexed; the forearms are semi-supinated, and there are no changes in the wrists and hands in relation to the forearms. The legs have begun the frog-like strokes. The thighs are abducted and completely outward rotated; the knees are semi-flexed, and the ankles are still plantar-flexed, the adjacent soles of the feet nearly facing each other. The Trapezii adduct, and the Latissimus dorsi, of both sides, complete the de-rotation of the shoulder-blades; the Latissimus dorsi bring the upper arms into adduction; the Biceps and Coraco-brachialis muscles flex the elbows; the Biceps and Supinator longus muscles semi-supinate the forearms, and appropriate muscles maintain the neutral position of wrists and hands. Abduction and outward rotation of the thighs use the Glutei and other intrinsic out-rotators; the hamstrings semi-flex the knees and the calf muscles keep the feet plantar-flexed.

[2] As a boy, the writer wondered if some sort of open and shut wooden flaps, with an open spread of perhaps twelve inches, and held in the palms of the hands, would not add to the effectiveness of hand paddling.

Count 3. *Arms shoot forward, knees are straightened and the feet far spread.*

The arm motions immediately continue those of Count 2. The shoulder-blades rotate upward; the forearms extend and abduct (but there is no rotation change at the shoulder-joints); the elbows fully extend, and there is no change in the forearm, wrist or hand positions, although fingers close and again form the prow-like V of the starting position. The lower limb movements, which attained the knee spread of Count 2, continue immediately by extension of the knees, by the use of the Quadriceps extensor of both sides. (Note: This phase, with the legs far spread and the arms pushing forward is the one of greatest retardation to the swimmer; but it is momentary and the powerful arm strokes, of Count 1, and the leg action, of the next count, greatly minimized the lessening of forward motion. However, because such slowing exists, most contest swimmers now use the crawl stroke, which has no such break in the momentum).

Count 4. *The arms are quiet. The extended legs snap together.*

The water-cutting of the now quiet arms, gives greater effect to the powerful leg action of this phase. There is strong adduction at the hip joints, with restoration to a neutral position of the outward rotation of the thighs of Count 3. It is presumed that the forward motion to the body is attained by the resistance of the water which is between the thighs. (It is interesting to note that the frog, whose swimming technique is, of the lower animals, nearest to that of the breast stroke in man, depends less upon his arm actions than upon his leg movements. The wedge shape of the frog's head lessens the water resistance).

Immediately after Count 4, with body in the position of the start, there is a short period (like that of the start) of "streamlined" glide, before the cycle begins again.

Throughout all phases of breast stroke swimming, there is complete lateral symmetry of muscle and joint mechanisms. It is therefore an ideal sport for the development of good posture.

The same may be said of other forms of swimming: The crawl stroke uses one-sided arm shoulder-girdle and body rotations, but these alternate with each stroke; swimming on the back may be either the same motions as the breast stroke, although with reversed body, or with alternate arm, shoulder-girdle and body twistings. The side stroke is strictly asymmetrical, as far as arm, shoulder-girdle and body activities are concerned; but if the swimmer more or less frequently gives each side the benefit of the change, this form is still advantageous. If, on the other hand, the side stroke swimmer's training habit conforms to using one side only, this method of swimming must be placed in a lower classification.

OTHER SPORTS PRESENTING LITTLE OR NO TENDENCY TO FAULTY BODY POISE

(Comments, but without analysis)

RUNNING is bilaterally symmetrical. Although preponderance of muscle effort is in the lower extremities, there is much activity in the trunk and arms.

BASKET BALL is a game of running, but the body bendings, the bi-symmetrical body twistings and the use of double-handed and largely overhead throwing, give it a high place in body poise. Indeed, Naismith,[3] the inventor of the game, had in mind the good postural effects of the high double-handed throw.

DIVING, using simple forward, back-dive, jack-knife, swan-dive or front or back somersault, is symmetrical in body action. If twisting dives are used and if the athlete uses right and left twists alternately, there is bi-lateral symmetry; but if his habit is to use a twist to one side to the exclusion of the other, he may, to that extent, develope body asymmetry.

SKATING is an even-sided body developer. There is greater stress upon the leg development, especially in race-skating; but the racer's forward crouch will prevent the bad hollow-back of faulty antero-posterior posture (see p. 170). On the other hand, the figure skater uses arms and body twistings; and as most figure-skating contests require bi-symmetrical performance of "stunts", figure-skating ranks high in good body poise.

SKIING, like skating, is bi-lateral in body activity, and the ski-jumper uses his trunk and usually "sails through the air" in a picturesque body poise.

[3] Naismith's publications of the time indicate this.

CANOEING, if there is frequent change of the paddle, from side to side, is an excellent back developer, and this development is evenly distributed on both sides. Canoeing does but little to develop the legs, although if the canoer "totes" his canoe on "carries" he is aware of much leg work.

SCULLING, a form of boat propulsion, using a sliding seat and out-rigged oar locks (like in Crew-rowing see below), but, in this case, using two oars, is an excellent symmetrical and well-nigh universal body developer.

ROW-BOATING, although lacking the leg development, gained when a sculler's sliding seat is used, is symmetrical for the trunk and arms, and is a good back-muscle developer.

HORSE-BACK RIDING uses many muscles in both legs and trunk, and is symmetrical. The sidesaddle, of our grandmothers, is decidedly asymmetrical and a menace to good posture.

RUNNING BROAD JUMP uses the entire mechanism of the legs in the run. In the take-off most athletes habitually use either one or other foot for the effort. The body and arms are effectively used in the jump. Except for the over-use of the take-off, this is a symmetrical action.

CHAPTER 7

II. SPORTS AND GAMES HAVING MILD TENDENCY TO FAULTY BODY POISE

CREW ROWING

The word rowing implies that act which propels a boat, by the use of oars, by one or more rowers seated in the boat, with back to the direction in which the boat moves. Sculling always implies one or two rowers who each use two oars. Race-rowing or crew-rowing refers to rowing in which two or more rowers, each using but one oar, take part. For increased power, the shell has a sliding seat, for each rower, to make available the added leverage of both legs. For symmetry of action on the boat or shell, an even number of rowers are involved. Thus two, four, six or eight oarsmen may be found in one shell. When there are as many as eight rowers, a coxman sits in the stern to steer the shell and to beat the rhythm of the stroke.

Rowing, in teams of greater or less number, is very old. From the times of Ancient Greece to the height of power of Mediaeval Venice, the war galley and the commercial galley were manned by rowers. These galleys had two or three banks of oarsmen, at different levels, on each side. They were the principal means of transportation of men and of goods. (It is interesting to note that in Lew Wallace's novel, Ben Hur, the hero, for the time being a galley slave, who saw the one-sided body development of his fellow rowers, prevailed upon the slave-driver to let him alternate, day by day, in rowing on the right and the left side of the galley.)

Crew rowing, as a competetive sport, has been largely a pastime of Anglo-Saxon peoples, and to this day has been, for the most part, amateur in character. In 1715, one Thomas Doggett posted a notice on London Bridge stating that he would give an emblem award to the winning six-man crew which should row annually, on the Thames "on the same day forever"; and it is said that, because of a provision in his will, it has been awarded ever since to the Fishmonger's Guild. The institution of the English Regatta, in 1775, started amateur racing crews from private club, from universities and from the great public schools of England. Organized crew racing in the United States started in 1811, when the "Knickerbocker" and the "Invincible" Clubs of New York raced on the Hudson River. In 1837, racing was transferred to Poughkeepsie, N. Y., but still participated in by private clubs. By 1852, the universities and colleges of the country became interested. Since then, although private clubs race annually, the principal crew racing interest is that among the universities and some of the larger preparatory schools.

THE KINESIOLOGY OF CREW ROWING (FIG. 92)

At the beginning of the stroke, the rower's feet are held in the foot cleats, the knees are bent, the sliding seat is far backward (in the shell), the trunk, neck and head are inclined far forward (backward in the shell), the arms are stretched, with elbows straight, far forward (backward in the shell), the hands grasping the oar handle, the oar handle is far backward, the oar slants forward and downward, passing through the swivelled row-lock and the blade of the oar is forward, ready to dab into the water. From that position, the entire mechanism—human and auxilliary—begins active motion. The ultimate power is the muscular mechanism of the rower, the levers are the many human segments involved, plus the sliding seat and the oar, and the ultimate work to be accomplished is the propelling of the shell forward in the water.

The levers are discussed as those of one rower. They begin at the most fixed point, the feet held in the cleats of the shell, known as the "stretchers".

Lever I—The bones of the legs (tibiae and fibulae).
    Fulcrum—the ankel joints;
    Power—the Soleus and posterior Tibial muscles, applied high up on the leg bones;
    Weight (resistance)—at the knees.
    Lever of the third order (FPW), moderately efficient as the power-arm distance is nearly equal to the weight-arm distance.

Lever II—The thigh bones (femora).
    Fulcrum—the knee joints;
    Power—the front thigh muscles (Quadriceps extensor group), applied high up on the thigh bones:
    Weight (resistance)—the pelvis and sliding seat.
    Lever of the third order (FPW), moderately efficient, as the power-arm is nearly equal to the weight-arm distance.

BODY POISE IN GAMES, SPORTS AND ATHLETICS 257

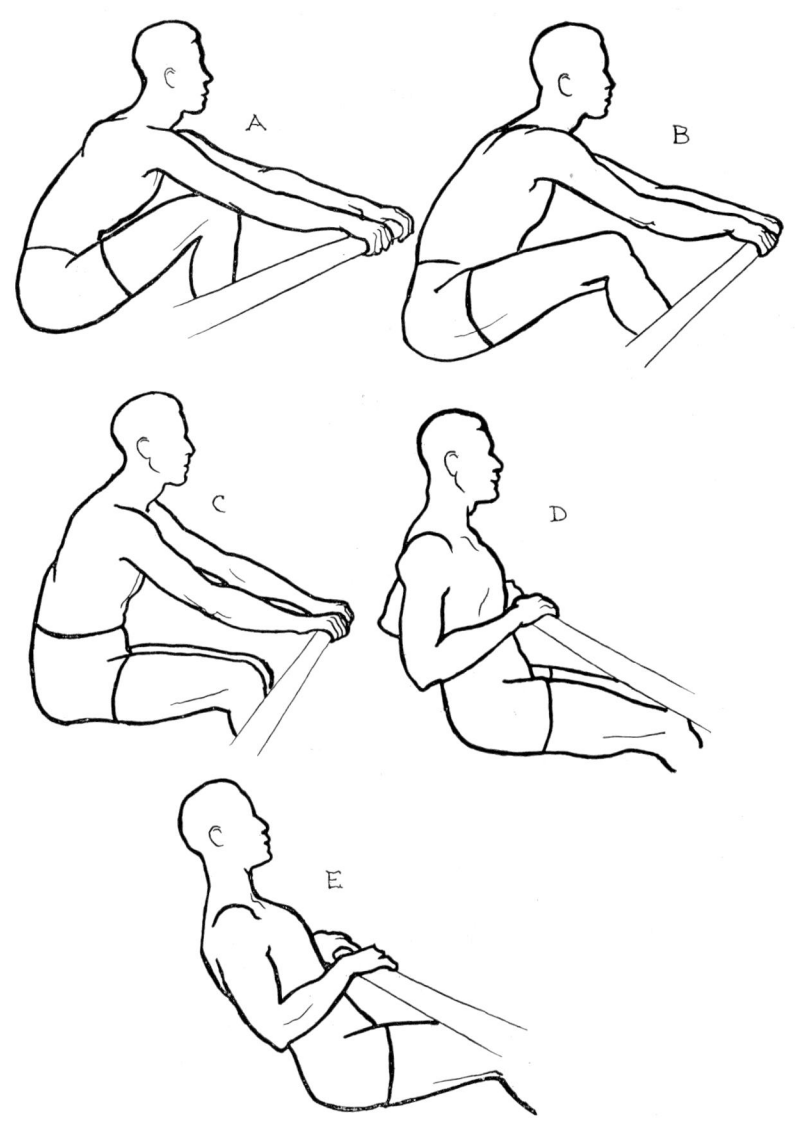

FIG. 92. Mechanism of crew rowing—five phases (redrawn from Nicholls and Mallan, Rowing, Pitman, London, 1939).

(Note: In levers I and II, comparative mechanical inefficiency is compensated for by relatively great power in the muscles involved. Also the principle of "resolution of forces", is not included in discussing these and other body levers, in order to avoid too great complication in this study.)

Lever III A—The pelvis, the spine and chest (and incidently the sliding seat).

Fulcrum—the hip joints;

(Note: As it is presumed that a spine, unbending in its segments except at the lumbo-sacral joint, is part of the training of "good form" in rowing, the many possible "breaks" in this long lever will be ignored; although the muscle training necessary to hold this comparatively rigid spine is advantageous in posture training. Also the complication of the chest, as a separate lever is here omitted, except to state that the development of increased chest measurements—chest capacity—increases the mechanical, as well as the vital, efficiency of the entire rowing mechanism.)

Power—the buttock muscles (Glutei), applied at the upper rear of the pelvis, and the back muscles (Erector spinae and allied muscles), applied at many points along the spine;

Weight (resistance)—the upper end of the sternum.

Lever of the third order (FPW), efficient because of the great strength attained by developed back muscles.

Lever III B—The trunk, the shoulder girdles and the upper ends of the upper arm bones.

The power, exerted by the back muscles, just noted, is greatly augmented by the strong Latissimus dorsi muscles, acting on the upper arm bones, when those bones have been extended (see Lever IV B, below).

The Fulcrum is the sacro-lumbar joint;

Power is applied on the upper arm bones (humeri) at about two inches below the shoulder joints;

Weight is the resistance at the elbow joints.

Lever of the third order (FPW).

Lever IV A—The shoulder girdle (clavicle or collar bone and scapula or shoulder blade, here considered as one to avoid complication) is a curved lever.

The Fulcrum is at the joint where the collar bones join the upper corners of the sternum (sterno-clavicular joints);

The Power is the Trapezii and allied muscles, applied at the inner borders of the shoulder blades, which pull the scapulae toward the spine;

The Weight (resistance) is at the shoulder joint, about midway between the fulcrum and the application of the power.

This is a lever of the second order (FWP) and is mechanically efficient, because the power-arm distance is about twice as great as the weight-arm distance.

Lever IV B—The upper arms (humeri).

Fulcrum—the shoulder joint;

Power—the strong Latissimus dorsi muscle attached on the upper arms, about two inches down;

(Note: The Latissimus dorsi also aids the leverage of the shoulder girdle, as well as that of the spine, as noted above.)

Weight—the resistance applied at the elbow joint.

Lever of the third order (FPW), mechanically not very efficient but actually a useful lever because of the great strength of and because of the direct (not oblique) pull of the Latissimus dorsi muscles.

Lever IV C—Forearms and hands.

Fulcrum—the elbow joint;

Power—the Biceps (and other elbow flexors), applied near the upper end of the forearm:

Weight (resistance) is that afforded by the oar handles.

Lever of the third order (FPW). Comparative mechanical inefficiency is counter-balanced by the developed strength of the Biceps and allied muscles.

(Note: The grip of the hands on the oar handles is important in changing the oar from "feather" to "dip" but is not an intrinsic part of the propelling mechanism of rowing.)

From this point, the leverage principles are those found in the oar and the shell.

* * * * *

Comparing this exercise with the discussion of the anatomy and kinesiology of Fundamental Standing Position (see Part I, Chapter 1), it is to be noted that all but one of the muscle groups essential to good posture are here used. They are the calf group, the

Quadriceps group, the back muscle group and the Latissimus group. The one group, necessary for the maintenance of good posture and not found in the boat-propelling portion of rowing, is that of the abdominal muscles. However, this muscle group is developed in the recovery part of the entire stroke—the trunk bending forward (backward in the shell). With this addition, it is evident that rowing is one of the best athletic exercises for developing the muscles essential to good posture.

Nicholls and Mallam,[1] in their excellent book on Rowing, go very thoroughly into all phases of the cycle of the rowing stroke, except the kinesiology. They stress the importance of all phases. They emphasize the action of the low back, of the position of the feet on the "stretchers" and of the position of the hands and especially of the wrists on the oar handles. On the muted question, among rowing coaches, of how far the trunk should sway backward at the end of the stroke, they make the following observations:

1. "The body swing in front of the perpendicular is what matters and not the swing behind the perpendicular"; and

2. "The effectiveness of the end of the stroke lies in the shoulder blades and shoulder-blade muscles, working the arms and the wrists and the hands. At the perpendicular trunk position, this shoulder blade action is most effective, because of the direct pull of all of the Trapezii or shoulder blade retractors, whereas at ever increasing backward position of the trunk, the shoulder-blade action is ever increasingly downward, on the arms and hands". (And so only the upper part of the Trapezii can act. WT.)

This writer feels that the reasoning concerning the shoulder-blade action is sound. He does not believe that the last word has been said concerning the lessened effectiveness of the trunk action, after the perpendicular has been reached. The authors quoted do not venture upon what backward trunk angle finish is to be used; but coaches in the United States are quite positive on the importance of this. They vary, however, between a finish at body perpendicular and thirty degrees backward.

[1] Nicholls and Mallan, Rowing, Pitman, London, 1939.

* * * * *

Crew rowing is so all-inclusive in its use of the body motor mechanism as to present no horizontal disproportions in relation to body poise. The momentary round-back of the bent-body start is neutralized by the strong shoulder retraction of the finish of the stroke. The excessive use of the back muscles is compensated, for abdominal muscle development, in the rapid recovery action from the finish to the start of the stroke. As to lateral asymmetry, in body development, the absolutely straight line of the sliding seat, in both active and recovery phases of the rowing cycle, insures perfect symmetry of action in both legs and in pelvis-hip controls; and, although the changing lateral angle of the oar causes some twist of the upper trunk and extra excursion of one shoulder blade more than another, the hands are placed so near together, on the oar, as to minimize this lateral asymmetry. That is, if a rower works habitually on a starboard oar, for instance, his upper spine will have a slight twist to the right and his left shoulder blade will be slightly more abducted, at the beginning of the stroke, than will the right; throughout the stroke, his upper trunk will untwist and his shoulder blades will have an equal degree of adduction, and finally at the end of the stroke, the upper spine will have had a slight twist to the left and his left scapula will have become more adducted than the right. (The further back from the horizontal the coach instructs for the end of the stroke, the greater will be these asymmetrical conditions of the upper spine and of the shoulder girdle.) Again, it is noted that these lateral asymmetries are at a minimum in crew rowing; but they do exist and they do give greater asymmetric development of the muscles involved—the to-left rotators of the upper spine (left Latissimus dorsi, right side External oblique of the abdomen, and the many local right-side rotators of the spine); also greater development of the left shoulder blade retractors (Trapezii and Rhomboids).

To meet this lateral asymmetry, the coach might take the hint from Ben Hur (see historical sketch) and, in the training period, reverse the positions of each oarsman daily, giving perhaps the last training week to the position desired for the race. It would completely overcome the body development tendency. And, as the writer firmly believes in the principle of absolute horizontal and lateral body development as the best preparation for specialized co-ordinated development and, indeed, for winning of contests, this alternation in side placing of the rowers is strongly suggested.

GOLF

Golf is an out-of-doors sport, in which one, two, three or four players follow specially made balls (one for each player) through a laid-out course, and using variously shaped clubs to propel the ball. Each "hole" consists of a teeing space, a fairway, a putting green and a sunken can, into which the ball must rest to complete the hole.

Holland was probably the original home of golf. The Dutch word, *kolf*, means a club. The Dutch often played on ice, with set-up sticks for the stakes. When it first appeared in Scotland is unknown, but by the middle of the Fifteenth Century its popularity seemed, in a measure, to have replaced archery. Some decades later, golf was played in England. There is record that, while playing the game on the links of Leith, King Charles I received the disturbing news of an Irish rebellion, but continued until he finished the game. The "Royal and Ancient" Golf Club of St. Andrews, in Scotland, founded in 1754, is still the most famous course in Great Brtain. Golf was scarcely known in the United States before the latter part of the Nineteenth Century. But since 1890, its popularity has spread, until no community of any size in America, is without one or more well laid out and maintained courses. Good golf courses are to be found today in France, in Switzerland, in Italy, in Canada, in Australia and indeed wherever sport-minded man locates.

KINESIOLOGY OF GOLF—THE DRIVE (FIG. 93)

Four phases of the drive, in a right-handed player, will be analyzed. These are: (1) addressing the ball—the stance; (2) the swing-back; (3) the drive, and (4) the follow-through.

1. *Addressing the Ball—The Stance*

The player stands to the left of the ball, the feet apart, an imaginary line connecting them being parallel to the line of direction the ball is to take. (Note: Some golfers prefer a slightly forward placement of the right foot, although many contend that any foot relationship, other than that of parallelism to the line of direction,

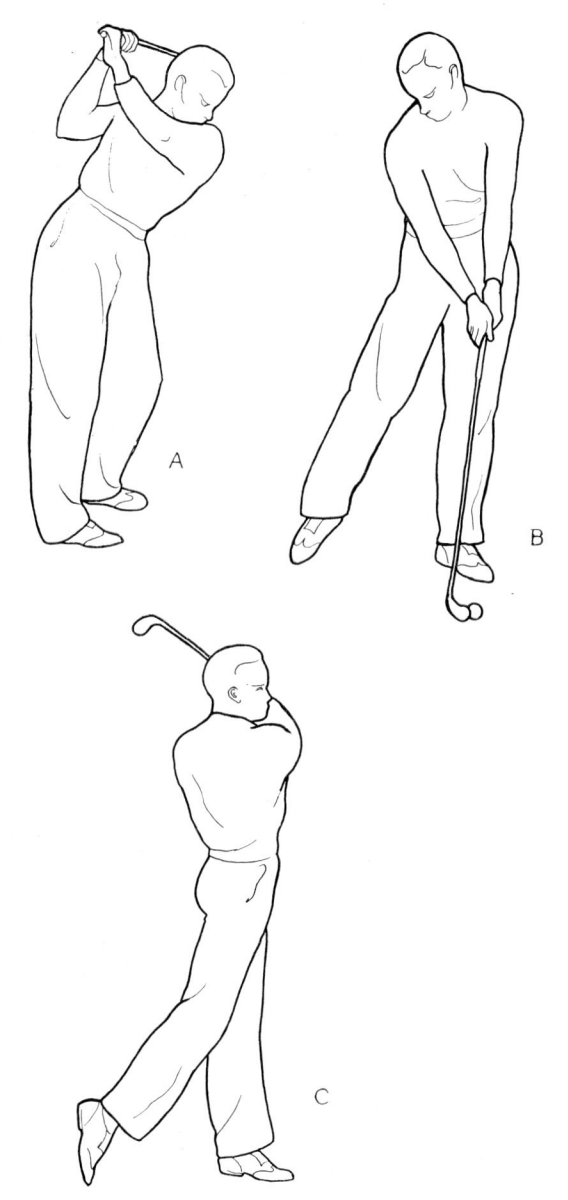

FIG. 93. Mechanism of golf, the drive—three phases (redrawn from Golfmasters, Golfmasters, Inc., N. Y.)

tends to divergence of the drive—a "draw", in the case of an advanced right foot, or a "slice" when the right foot is rearward.) The heel of the left foot is so placed that a line connecting it with the ball would be at right angles to the line of direction of the ball. The knees are fully extended; the thighs are abducted and the pelvis flexed on the thighs; the pelvis is laterally parallel with the line of direction of the ball; the spine bends with the pelvis—no more and no less—, that is, although the entire trunk is slightly forward-bent, the relation of its segments, to the pelvis and each to each, is not other than is found in normal good posture—; the head and neck are enough forward-bent, on the trunk, to insure that the eye may accurately watch the ball; the shoulder-blades are moderately abducted; the arms flexed at the shoulder-joints; the elbows are fully extended; the wrists are in slight ulnar flexion, and the interlocking hands are flexed in the grasp on the handle of the driver, the right hand just in front of the left hand. (Note: The wise golfer chooses such a length of club as shall allow the straight arms, when the trunk is inclined forward about thirty degrees, in addressing the ball and at the moment of contact of the driver-head with ball, at the end of the drive.)

This is a position of readiness for the movements which follow, rather than one of active effort. The muscle-joint mechanism of the forward-bent pelvis (with accompanying trunk) involves eccentric contraction of the Glutei muscles, and, to maintain the forward position of the trunk, the Erector spinae muscles of both sides are in tonic contraction. The Extensor muscles of the neck, on both sides, are in eccentric contraction; the shoulder-blade adductors are in eccentric contraction; the arm-flexor muscles (shoulder-joint flexors) are in concentric contraction, and all finger and thumb flexors of both hands are in concentric contraction.

2. *The Swing-back*

As the club is raised from the position of addressing the ball and carried to the right side upward to back of the neck, the foot stance remains unchanged, except for a slight raise of the left heel, as the left knee bends a trifle. Otherwise, the stance of the thighs and legs and feet remain unchanged.

(Note: Some golfers, while practicing the slight left knee bend, debar any left heel-raising.) The most characteristic features of good form, in the mechanism of the swing-back and of the drive to the moment of striking the ball, are: (1) the maintenance of the pelvis and trunk-bend and especially (2) the rotation of the entire trunk on (and at no time away from) an imaginary axis which passes from the top of the head, through the trunk and which emerges at the lower tip of the spine (coccyx). This axis line inclines about thirty degrees from the vertical and (if it were ground-plane projected) is at right angles to the line of direction the ball is to take. On this axis, the pelvis twists (in the swing-back) to the right and the spine and trunk increasingly twist, from below upward, to the right, until the transverse of the shoulder-girdle is at about a right angle to the line of direction of the ball. The right scapula adducts and rotates upward; the right upper arm is drawn to the side of the body and a little backward; the right elbow fully flexes and the position of the right wrist and hand remain unchanged, except for moderate radial flexion of the right wrist. At the same time, the left shoulder-blade abducts, the left upper arm adducts, as it is carried upward across the front of the chest, the left elbow flexes and the left wrist flexes somewhat. The head remains forward bent, but now in a to-the-left-twisted position, in relation to the trunk, so that eyes may continue to watch the ball by glancing over the left shoulder.

This movement involves the use of the calf muscle (Gastrocnemius) to raise the left foot; eccentric action of the left Quadriceps extensor group, and of left Gluteus, to maintain left hip-knee bend; concentric action of right hip out-rotators and of left hip in-rotators, to twist the pelvis to the right; concentric action of right Latissimus dorsi, of left External oblique and right Internal oblique muscles of the abdomen and of the multiform rotators proper of the spine, to further twist the trunk to the right; of right Trapezius and Serratus magnus, to adduct and upward rotate the right shoulder-blade; of right Pectoralis major and of right Latissimus dorsi, to hold the upper arm to the side of the chest and backward; of right Biceps cubiti, to flex the right elbow, and of all right wrist steadiers and finger flexors, to maintain the grasp on the club; and finally of left Pectorals, to abduct the left shoulder and left upper arm, and the left elbow and wrist and finger controlling muscles aid in holding the club at its elevated and backward placed position behind the neck. There is eccentric action of the Extensor muscles and concentric action of to-left rotating muscles of the head, to maintain the "watch the ball" position of the head.

This analysis of the joint-muscle mechanism involves a considerable asymmetry of the trunk, the shoulder-girdle and of the arms.

3. The Drive

The drive in golf is that portion of body activity which occurs from the end of the swing-back to the moment of contact of the club-head with the ball. It reverses much of the human mechanism, especially the asymmetry of the trunk, of the shoulder-girdle and of the arms. But the maintenance of the trunk *on* its rotating axis (see above) is not changed until the follow-through begins. The relative position of the feet on the ground continues, but as the club nears the ball, the left hip-knee bend is straightened. The to-right-twisted pelvis untwists; the to-right-twisted trunk also untwists, but does not lose its forward-bent position. The left shoulder-blade adducts; left upper arm abducts and remains somewhat flexed; the left elbow and wrist are straightened, and the left hand retains its grip on the club. The right shoulder-blade abducts and de-rotates; the right upper arm adducts and flexes to thirty degrees forward; the right elbow is straightened, the forearm semi-supinated and the right hand-grasp is maintained on the club, just below the left hand-grasp.

The movements of the drive phase, in golf, involve concentric contraction of front thigh muscles (Quadriceps), to straighten the left knee; of right Glutei and left Ilio psoas muscles, to straighten the to-right twisted pelvis; of left Latissimus dorsi, right External and left Internal oblique muscles, of the abdomen, and of the many rotators of the spine, to untwist the trunk. The mechanism of the left shoulder-girdle is motivated by the Trapezius muscle, which adducts the shoulder-blade, and by the Pectoralis muscle, which abducts and rotates the right shoulder-blade downward.

Abduction and thirty degrees flexion of the left upper arm is caused by the rear portion of the Deltoid and by the Biceps (long head) and the Coraco-brachialis muscles, and the left elbow is straightened by using the Triceps muscle. The adduction and thirty degrees flexion forward of the right upper arm uses the Pectoralis major, the front portion of the Deltoid and the Biceps (long head) and the Coraco-brachialis muscles, and the right elbow is straightened by the Triceps and the forearm is semi-supinated by Supinator longus and brevis muscles. Firmness of both hand grasps, accompanied by flexibility of both wrist actions, are accomplished by the many muscles controlling these parts. (Note: Although the muscle mechanism of the body twist, of the shoulder-girdle play and of the upper arms provides the main power to the drive, the nicety of right wrist muscle co-ordination give accuracy.) While all of this activity takes place in the mechanism below the neck, the movements of the head and neck are important. The player must "keep his eye on the ball" clear through and to a trifle after the driver contacts with the ball. The to-left twist of the head, at the top of the swing-back, is untwisted, *pari passu*, with the untwist of the trunk during the drive. This requires the use of the to-right twisting muscles of the head and neck, which action ceases when the head faces forward in relation to the chest.

4. *The Follow-through*

This is the action of the player from the moment of the contact of the driver-head with the ball to the end of carriage of the club far forward and upward. It involves no change in position of the feet upon the ground, except that the right heel may rise slightly and the right knee may bend. The right toes act as a guiding pivot while the pelvis (carrying the right unflexed thigh with it) and the trunk continue the untwisting, of the drive to a definite twist to the left. That is, the right side of the pelvis and of the body are carried forward. The left hip joint, stabilized on a practically immovable left limb, is the principal fulcrum for this twisting action. The swing of the club is a definite continuation of that of the drive, and describes a great circle, defined by the head of the club, to perhaps ninety degrees more than the three hundred sixty degrees of a circle. Its purpose is to prevent any "let-up" in the force and in the accuracy of the drive, just as the race horse puts his entire effort into crossing of the finish line, which sustaining of speed makes it necessary for the horse to run some yards past the finish. After the driver-head has definitely hit the ball, the forward-bent trunk is raised and the eyes may stop looking at the vacant peg, in order to observe, with self-congratulation or otherwise, the direction the ball is taking. (If "otherwise", he hopes he has an alert caddy!)

The use of other golf clubs—fourteen in all are allowed in tournament matches—involves body mechanisms similar to that of the driver, though each club requires its own body mechanism.

In studying *golf in relation to body poise*, one notes that, although the swing-back to the right is accompanied by a decided asymmetry of the trunk, shoulder-girdle, arms, head and neck mechanisms, the drive and the follow-through completely reverses these asymmetries. However, the range of the many asymmetrical joint excursions of the drive and the follow-through is greater than the range of the opposite joint excursions of the swing-back, and the greater muscular effort, and therefore gradual preponderance of muscle development, of the drive and follow-through become possible factors, in the formation of scoliosis, which should be considered. Golf must therefore be placed in the second group of the classification of this book.

The use, by the golf devotee, of the "key-note" exercise, described in this book (see p. 228), should *obviate* any such *harmful tendency*. Of course, the three-mile walk, necessary to cover eighteen holes, is a healthful feature of the sport.

Although not within the scope of this book, it is nevertheless in order to call attention to the real heart danger, which may follow the over indulgence, in golf, by elderly people, who otherwise lead sedentary lives, which they perhaps too thoughtlessly change to the vigors of the game.

BOXING

Boxing is described as the "art of attack and defense with the fists protected by padded gloves". This is in contrast with pugilism, in which bare fists were used.

The Ancient Greeks indulged in boxing matches. The English have supported boxing, both amateur and professional, for more than two hundred years. Jack Broughton (1705–89) was called "the father of British pugilism." He invented and first used the modern boxing-glove. John Jackson, called "Gentleman Jackson", was the popular teacher of boxing. He had many pupils of rank, among whom was Lord Byron. In 1866, the Amateur Athletic Club was founded, and one of its members, John G. Chambers, with the help of the Eighth Marquess of Queensbury, drew up a code of procedure known since then as the Queensbury Rules. In America, boxing was indulged in, in an unregulated and unscientific way, until the Amateur Boxing Association, founded in 1884, formulated rules of procedure. Boxing in France dates from about 1830.

Professional boxing in America has included the names of men, some of whom practiced boxing on a high plane of sportsmanship and some who did not. Among those who have been outstanding for clean sport may be mentioned James J. Corbett, Gene Tunney and Joe Louis. Amateur boxing, in America, has had many followers in universities and colleges and in private clubs.

KINESIOLOGY OF BOXING

The physical activities of boxing are too diversified to allow detailed kinesiological analysis; but some generalizations are possible. As the body segment activities are so varied and important in the sport, the base of support is of prime interest. This changes from moment to moment, but it must always be firm, both for offensive activity and for defense against the upsetting effects of hard hits by the opponent. The relative positions of the contestants on starting may be described as follows: "The boxers face each other just out of reach and balanced equally on both feet, the left foot from ten to twenty inches in advance of the right".[2] From this start, although the boxer is constantly stepping or even hopping in any direction that attack or defense may require, he always endeavors to keep a large base of support. The knees, for the most part, are kept fully extended or, if the action requires, one or other or both may be slightly bent. The hip joint positions vary from slight flexion to extension, from inrotation to outrotation, from abduction to adduction, but they seldom go through anything like their anatomical limits of activity. The muscular co-ordination of hip, knee and ankle-foot controls are well trained and must be under the conscious or unconscious regulation of the

[2] Encyclopaedia Brittanica.

fighter, while in action. The pelvis is subject to the same great variation of position and of muscular control, but rarely changes much in any direction from that of Fundamental Standing Position.

But when the parts above the pelvis are studied, an infinite variety of movements of trunk, of head and neck and of arms are noted. There are flexions and extensions and twistings and side motions of the spine, and any of these may extend to the full anatomical limit. The co-ordinated muscle controls are, of course, those of the individual joint controls involved. They are multiple joint actions. Some boxers practice the so-called "crouch" for offensive action. This is a forward bending of the trunk, with somewhat increased bending of one or both knees and hip joints. It involves flexion with or without twist or ride motion of the lumbar spine. The users of this position believe that it has strategic value, especially in the offensive, but from the point of view of body mechanics it has lessened universal adaptability in the offensive than has the upright or more nearly upright position, and to the writer it would seem to increase the hazards of the defensive, unless perchance the user of the position has a head of "ivory". The head and neck are never in the offensive, in boxing. In the defensive, the neck must be facile, with the use of any of its anatomical movements and employing the complicated muscle controls of the region, in dodging the blows of the opponent. Injuries to the face are ever present risks to be guarded against.

It is especially in the description of the activities of the shoulder girdle and of the arms that one is confronted with the well-nigh impossibility of presenting detail. Probably no other sport or athletic activity involves such diversified movements in the upper extremities as occurs in boxing. There are adductions and abductions, elevations and depressions, upward and downward rotations of the shoulder blade, with corresponding movements of the collar bone; there are flexions and extensions, adductions and abductions, rotations inward and outward and circumductions at the shoulder joint. Here are used flexions and extensions of the elbow joint, pronations and supinations of the forearm and flexions and extensions and lateral motions of the wrist. And all of the many activities are controlled by their appropriate muscles. Only in the fingers, which are glove-encased in the flexed position, is there but little activity, although an opened hand, with an extended wrist, is sometimes legitimately used to push away an opponent. But, in general, it may be stated that in boxing there is more *intensive* forward than backward movement of the shoulder girdle and of the arms. That is, there is more abduction than adduction of the shoulder blades, requiring more development of the Pectorals than of the Trapezii and the Latissimus dorsi muscles; more flexion than extension at the shoulder joint, necessitating more development of the Pectoral and of the front of the Deltoid muscles than of the rear of the Deltoid and of the Latissimus dorsi muscles, and more extension than flexion at the elbow joint, requiring more development of the Triceps than of the Biceps and of the Coraco-brachialis muscles. The great development of the Pectoralis major and of the Triceps are notable features of the boxer.

Viewing boxing in *relation* to good *body poise*, one would give it a fairly high place. There is nothing in its pursuit which tends to lateral body asymmetry. Most boxers, although they (or their adherents) talk of the "right punch" or the "left punch", do and indeed must develop bilateral hitting. Also, although the intensive activity is in the upper segment of the body more than in the lower, the "skipping about" in the ring, of most boxers requires and uses enough muscle co-ordinations of the legs to minimize the tendencies of "horizontal disproportions", outlined in the introduction to this part of the book. However, the overdevelopment of the Pectoral muscles, to the disadvantage of the upper back muscles, does tend to round shoulders, and so removes boxing from a first place in body poise.

To *meet* this round shoulder tendency necessitates strict adherence to the general body training, which the best coaches give to the professional boxers, and the adoption of good setting-up drills, by college and athletic club physical directors and coaches, and the indulgence by the contestant in the counteracting use of such sports as rowing and swimming. Some of the prone-lying corrective exercises, such as the "seal", found in Chapter 4 of this book, are excellent counter-actors of the round shoulders tendency of boxing.

OTHER SPORTS PRESENTING MILD TENDENCY TO FAULTY BODY POISE

(Comments, but without analysis)

FOOT BALL is classed as an all-around body developer. There is much running, trunk bending and body twisting. In the "open game", developed some years ago, there is one-sided throwing, and there has always been single-legged booting of the ball. These asymmetries usually affect but few of the players.

FIELD and ICE HOCKEY are more or less universal body developers, with perhaps preponderance of leg work. But the body-twistings and shoulder-girdle activities are marked. The use of the stick, in one hand more than the other, tends to asymmetry. However, as a good player often uses both hands and not infrequently changes the stick to the opposite hand, one-sidedness is minimized.

TENNIS, although definitely using one arm and one-sided trunk-twist more than the other in the "serve", nevertheless involves bi-symmetrical trunk-twistings and use of the legs, in continuing play, after the serve. The tendency to faulty posture is not great.

BICYCLING uses the legs more than the trunk or the arms. In the crouch of the racer, the forward bent trunk serves to prevent hollow-back (lordosis). In classification of its relationship to good posture, it is to be considered as "horizontal disproportion" (see p. 241).

FENCING has an historical background. It has much in its favor in relation to good posture, especially of the trunk and head and because abduction of both shoulder-blades is predominately used. But the excessive use of the right arm and shoulder mechanism places the art in this group.

Chapter 8

III. SPORTS AND GAMES WITH GREATER TENDENCY TO HORIZONTAL DISPROPORTION OF MUSCULAR AND STRUCTURAL DEVELOPMENT AND LATERAL ASYMMETRY.

BASEBALL

Baseball is a typical and leading American game, although the older English game of Rounders bears some similarity. The early American games Town Ball and One Old Cat, Two Old Cat, Three Old Cat and Four Old Cat also show some relationship. But baseball, as we know it today, is a specific American game, played between two teams, with definite rules and on a prescribed field.

Most students of the history of the game acknowledge Col. Abner Doubleday as the real founder of the game, as it is played today. In 1839, he devised the diagram of the field, the position and distance of the bases and the position of the players. From time to time, since then, rules of the game have been formulated. Since the early part of the Nineteenth Century, the game has spread from city to city, from village to village and into the schools and colleges. At first it was entirely played by amateurs. Gradualy semi-professionalism (or the employment for pay of outstanding players to augment the efficiency of amateur teams) and complete professionalism (or use of an entire team of paid players) appeared. Complete professionalism started in and around New York City and spread from city to city and from village to village, so that today there are many leagues with stated seasonal schedules. The National League, founded in 1876, and the American League, started in 1900, constitute the so-called Major Leagues. They each play series of games to establish relative percentage ratings of victories and especially to determine the respective league championships; and at the end of the season, a shorter series of games is played between the two champions to determine the World's Championship. Throughout the United States and in Canada there are many other leagues or associations, called the Minors, which also play their individual series and, to some extent, finish the season with inter-league games.

Meanwhile amateur baseball has not lagged in the United States and Canada. There is hardly a town or village, a school or a college that does not have its amateur baseball club.

Some years ago, largely through the influence of postseasonal trips of one or other professional team, which toured the world, baseball has made some headway in England and Ireland, but especially in Japan and the Philippine Islands. Indeed, it was said, in the first trying days of American occupation of the Philippines, that "England ruled her colonies with the dinner coat, America with the baseball bat".

THE KINESIOLOGY OF PITCHING

Three phases of the over-hand throw of a right-handed pitcher will be analyzed—the "wind-up", the course of the movement, and the finish.

A. *The Wind-up* (Fig. 94A). The right foot is firmly placed on the ground at 90 degrees or more outward from the line of direction which the ball is to take. This requires firm tonic contraction of all ankle-foot control muscles. The right knee is bent 20 to 30 degrees, necessitating contraction of the right Quadriceps extensor muscles. The right hip joint is flexed and abducted, with strong contraction of the right Glutei muscles and general tonic contraction of all other stabilizing hip-joint muscles. The pelvis is tilted upward and forward. The entire left limb is raised in front in preparation for the forward step. This uses the left hip flexors, the left Quadriceps extensor and the left ankle dorsi-flexors. The trunk tilts backward but without twist on the pelvis; but as the pelvis is twisted forward to the left (in relation to the right limb), the trunk has a relative left forward twist. This uses the left Erector spinae, Quadratus lumborum, left abdominal muscles, to attain the trunk position, and the opposite muscles to hold it. The head and neck are twisted and bent to the left, using the many muscles connecting the upper trunk to the cervical spines, the cervical spines to each other and the cervical spines to the head. The right arm, carrying the ball, is slightly abducted at the shoulder joint, the elbow slightly flexed and the wrists slightly dorsi-flexed. (The position of the forearm, wrist and hand varies somewhat according to the pitcher's intent to throw a straight, an in- or out-curve or a drop ball.) The muscles controlling the right arm need not be specifically named but it is to be noted that they are in tense contraction, in readiness for the throw. The left arm is carried in abduction of about 90 degrees

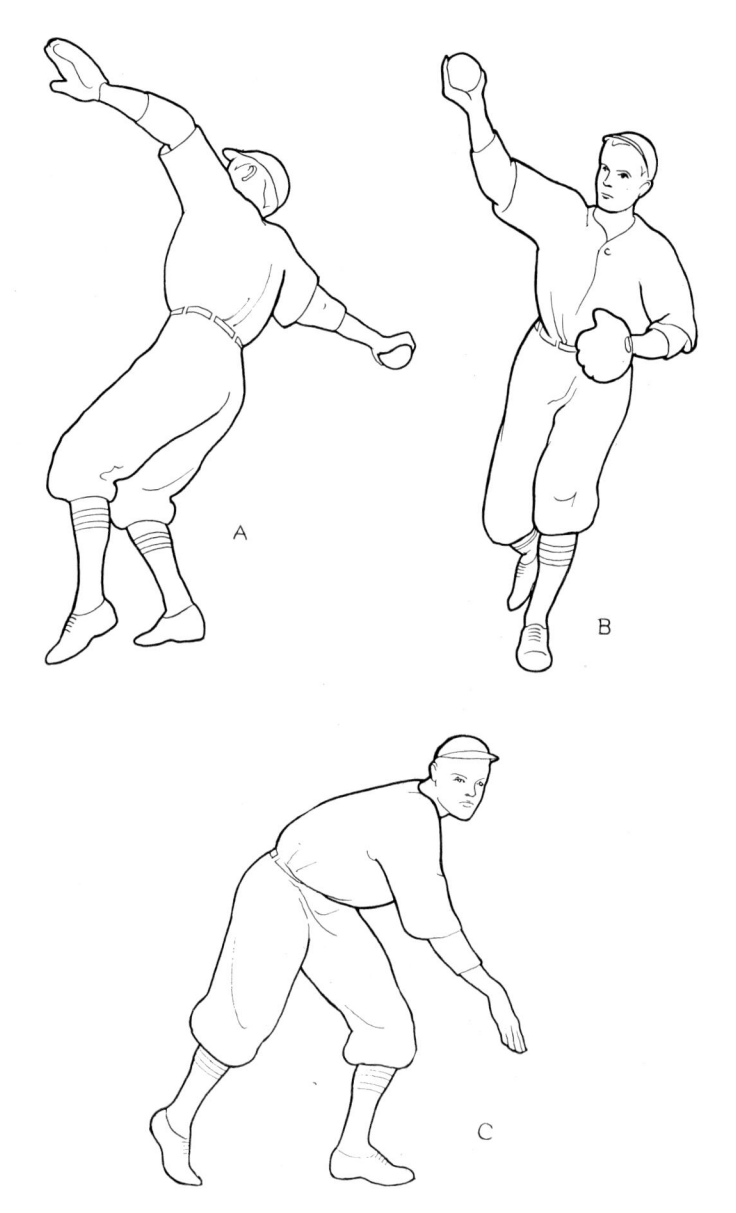

Fig. 94. Mechanism of baseball, the pitcher—three phases (redrawn from Ethan Allen, Major League Baseball, Macmillan, N. Y.).

and in inward rotation at the shoulder joint and in full extension of the elbow and the wrist. This position of the left arm is purely one of maintaining the entire body balance of the wind-up position. This wind-up position, although but momentary, is nevertheless one of stable equilibrium, the line of gravity passing somewhat to the rear of the head, to the right of the mid spine, through space to the knee joint and to the front of the heel.

B. *The course of the movement* (Fig. 94 B) from the wind-up to the finish, or delivery of the ball, is one in which the entire length of the body plus (at one moment) the vertically raised right arm is used as a very effective lever for the delivery of the ball. This movement consists in a plantar flexor and adduction of the right leg on the right ankle, using the plantar flexors and adductors of the ankle; a complete extension of the right knee, using the right Quadriceps extensor muscle; of extension and outward rotation at the right hip joint, using the right Glutei muscles and the adductors of the left hip, first to level the pelvis laterally, then to bring the pelvis forward and to lower it on the right side. The trunk twists markedly to the right and is carried forward, *pari passu*, with the pelvis, using first the left Latissimus dorsi, the right External oblique and the left Internal oblique muscles of the abdomen and the many smaller rotators of the lumbar and thoracic spine. These muscle controls constitute the initiative of the important trunk activity, but after the trunk has reached the vertical and tends to the downward position, the muscle action just named continues, and the muscles opposite to them are in eccentric contraction to hold the trunk in the pendant position. To keep the head forward-facing, the head twists to the right, using the appropriate rotator muscles of the neck. In the course of the movement the entire right arm mechanism takes a straight arm circuit of about 270 degrees, to the point of delivery of the ball, and adds about 90 degrees of "follow-through" to the position of the finish, and the shoulder girdle moves from the anatomically neutral relationship to the chest, formed at the "wind-up", to marked shoulder-blade abduction at the finish. This involves, in the delivery of a "straight" ball, strong contraction of the right Pectoral muscles, of the right Deltoid, of inrotators at the shoulder joint, of the pronators of the forearm and finally of flexion of the wrist and opening the fingers. (The mechanism of delivery of curved balls includes variations of movements of the right forearm, of the wrist and especially of the sequence in opening of the fingers and the thumb, with possible momentary backward jerks of the shoulder girdle. It is too individualistic for kinesiological analysis here.) The movement of the left arm, which is mostly an action to maintain balance, is a straight-armed one from forward and somewhat upward, sideways and somewhat downward to a backward and horizontal position, going through an arc of perhaps 135 degrees. It is activated by the entire left Deltoid muscle and by the left Latissimus dorsi, the muscles controlling the elbow and the wrist being in tonic contraction. The line of gravity changes from moment to moment throughout this movement.

C. *The finish* (Fig. 94C) presents a base of support described by the well-planted left foot and the tip-toes of the right foot. The left foot faces forward and a trifle outward. The left leg is in practically neutral relationship to the foot, using all ankle stabilizing muscles; the left knee is about one-quarter bent, the left Quadriceps muscles being in eccentric contraction; the left hip-joint is about half bent and, as the pelvis is carried downward and to-right twisted, the left Glutei are in eccentric contraction and the left hip adductors are in concentric contraction; the trunk is carried forward to the right, with the pelvis, using the left Erector spinae and associated muscles to hold the position. The head is twisted to the right, and forward, in relation to the upper trunk, using the rear muscles of the neck to hold this; the right arm is downward and backward, involving the right Pectoral muscles principally, and the left arm is horizontally backward, using the left Trapezius, Deltoid and Triceps muscles.

Kinseiological analysis of the batter, of the catcher and of other players, interesting as they are, will not be included in this book. The pitcher uses by far the most energy of any member of the team. Indeed, in professional baseball, the games of which are played almost daily throughout the season, rarely is any one pitcher used oftener than once in four successive games.

Baseball is a valuable game, but it must be classed in the third group of this section. It presents little or no element of horizontal body disproportion, but it definitely *tends* to *lateral asymmetry*, due to the excessive use of one arm and shoulder girdle more than the other and of the unilateral body twists accompanying the arm motions.

What is to be done about it? Excellence in accomplishment is based on acquired unilateral skills. Where, as is not infrequently found, a right-handed pitcher or thrower is a left-handed batter, and vice versa, the one-sided effect is largely neutralized. It is not reasonable to ask professionals or members of teams representing schools and colleges to develop contra-sided actions, but the suggestion is here made that, within the scope of intra-mural athletic activities, the physical director or coach may plan some games in which the teams must use the side of the body opposite to that of their usual habit. But, of course, the best *remedy* to the *faulty body tendencies* of baseball is the pursuit of stated gymnasium calesthenics or setting-up drills or in participation in such sports as rowing, swimming and walking. Individuals may practice in private, or under proper instruction, going through their usual motions, in pitching or throwing or batting, but with the contra-symmetrical body mechanism, and with care to simulate the action of the usual side as nearly as possible. The writer believes that this procedure would indeed tend to improve the results, when returning to the usual activities. *Concomitant symmetrical activity always conduces to better unilateral activity.*

POLE-VAULTING

Pole-vaulting is a jump over a horizontal bar, which may be placed at more than twice the height of the jumper, who uses a long bamboo pole to clear the bar. From the firm hand-grasp on the pole, the vaulter makes vigorous body-swing and shoulder and arm heave upward, to attain a high level.

Primitive man used the long pole, not so much to attain heights as to clear streams and other obstacles. In modern times, pole-vaulting is an event in track meets. Up to the eighteen nineties, the records of this sport were held in the town of Ulverston, in Lancashire. The athletes of this English community evolved and used a method of clearing the bar which might be called "pole climbing". The alternate hands rapidly grasped the pole higher and higher, as the body-swing, with the aid of the single-arm heaves, pulled the vaulter to more and more elevated positions. This method has now been barred. Since the use of the bamboo pole and the fixed one-hand grip, with but one upward slide and re-grip of the other hand, has been the rule, the event has reached records of ever-increasing height. A bar clearance of approximately fifteen feet has been attained. This sport is now a spectacular part of most college and inter-collegiate meets in the United States, in great Britain and in the Scandinavian Countries.

KINESIOLOGY OF POLE-VAULTING (FIGS. 95 AND 96)

Two methods are used: (a) the swing of the legs forward upward, with shoulder-girdle heave, before the pole is vertical, the body turning to face downward, while clearing the bar; and (b) the swing of the body sideways upward in one action. The first method will be analyzed. Herein, the grip and carry, the run-up or approach, the clearance, the push-away and the landing will be briefly described. The phases of the pole-plant and the take-off, of the swing-up and the heave, and of the body-twist will be kinesiologically analyzed.

1. *The Grip and Carry* (Fig. 95A). The vaulter, having previously paced off his running distance and having estimated the height on the pole at which his right hand-grasp is placed (a position which he does not change until the final push-away), stands for a moment at the start of his run-way. His right hand is well back of his hips (Fig. 95A), his left hand-grasp is forward, with elbow slightly flexed, so as to carry the pole forward. His left foot may be off the ground, with his right hip and knee slightly bent and the limb raised, as he begins the run.

2. *The Run-up or Approach* (Fig. 95B). The speed used in the approach is a matter of individual experience. Its purpose is to acquire just as much momentum as is consistent with the abrupt change of direction, which change of direction occurs as soon as the end of the pole is ground-fixed and the take-off "stamp" is made. During the run-up, the pole is raised, from the hips level, to high over head and forward. As the end of the pole points down (aiming at the take-off box), the left hand slides upward, on

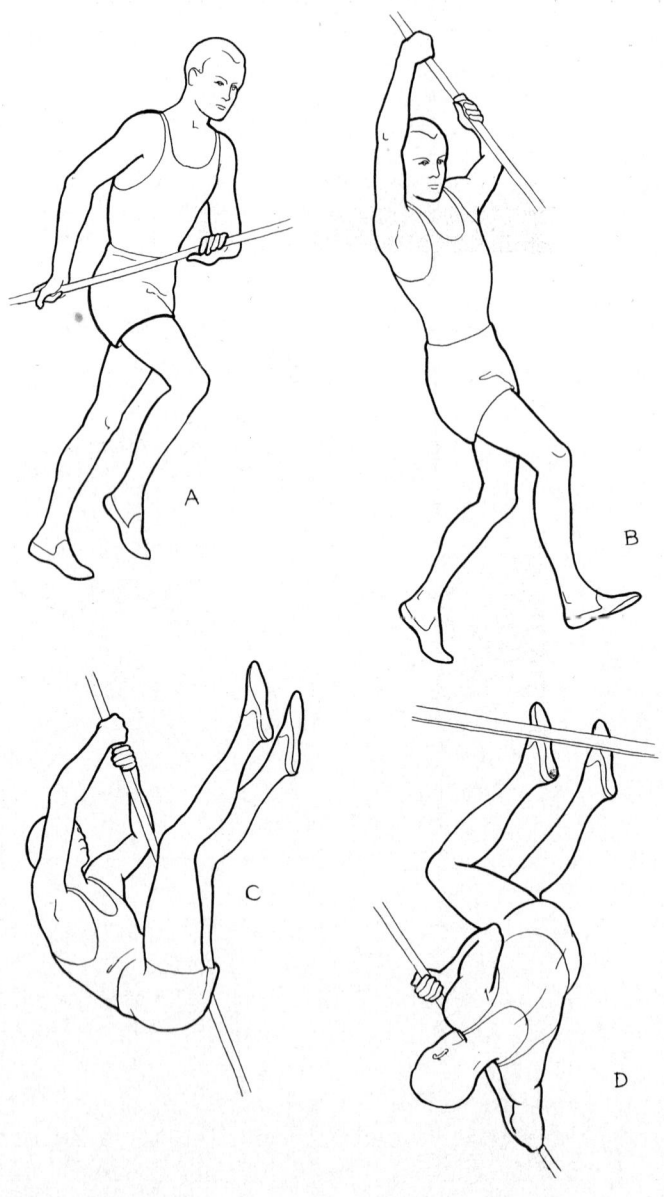

FIG. 95. Mechanism of pole vaulting—four phases (these illustrations are based upon the illustrations from Conger's Track and Field, copyright 1939 by A. S. Barnes & Co.)

Fig. 96. Mechanism of pole vaulting—three additional phases

the pole, to just below the fixed position of the right hand, and the raised right foot is ready for the take-off stamp.

3. *The Pole-plant and the Take-off* (Fig. 95C). The end of the pole is fixed in the box; the arms, nearly parallel, have thrust the pole forcibly down, using both Pectorals and both Latissimus dorsi muscles; the right limb is forced strongly downward, using right Glutei and right Quadriceps extensor muscles, to attain the right foot stamp, which is an important feature of the upward body-swing.

4. *The Swing-up and the Heave* (Fig. 95D). From this point to the final landing in the pit, the vaulter is free from the ground. The pole is his means of support until just before the final push-off, but there is much change in the relationships of his body to the pole. The hands grip the pole firmly, the nearly parallel limbs are first flexed and carried close to the right of the pole and forward upward, using all hip flexors of both sides—the Iliopsoas, the Rectus femoris, etc. Then the heave begins. Both arm and shoulder-girdle mechanisms draw first the upper trunk and then the entire body upward. At this phase, the parallel legs, half-bent at the hips, lead the entire body into a position nearly upside down. To accomplish the heave, the elbows flex, using the Biceps cubiti and associated muscles; the shoulder joints extend, employing the Latissimus dorsi and Triceps muscles; the shoulder-blades adduct bringing the Trapezii muscles into use; the trunk is carried upward on the shoulder-joint fulcrum, requiring first the use of both Pectoralis major and minor muscles, and then the abdominal muscles, to raise the lower trunk and the legs. When a reverse body-slant upward is reached, both limbs are nearly fully extended.

(The body-twist starts from the beginning of the up-swing and the heave but will not be described until the next phase.)

5. *The Body-twist* (Fig. 96E). The previous phase, the up-swing and the heave, has brought the body of the vaulter to nearly the highest level he attains in the effort, and the body twist to the left has already begun. The grip of the right hand on the pole is the essential fulcrum, the distance on the pole to the left hand-grip is the lever-arm distance and the extending left elbow and shoulder-joint and abducting left shoulder-blade, constitute the power application. This power is supplied by the left Triceps

and by the left Pectoral muscles. At the same time, the right elbow is flexing, the right shoulder-joint is extending and the right shoulder-blade adducting. This requires the use of the right Biceps and Coraco-brachialis and of the right Latissimus dorsi muscles, to attain the twist of the upper trunk in relation to the shoulder-girdle, the arms and the pole. For the left-twist of the lower trunk, the pelvis and the parallel legs, the lower border of the chest and the thoracic spine must be considered as the broadly extended fulcrum. The varying distances from the front to the rear of the pelvis and the lengths of the spinous and transverse processes of the lumbar spine constitute the lever-arm distance. The abdominal and many rotator muscles of the lumbar spine constitute the leverage power. The right External oblique and the left Internal oblique are the abdominal muscles used and the multiform rotators of the spine, some on the right and some on the left side, definitely aid in the action. It is to be noted that many pole vaulters also use a "scissors kick," when nearing the bar, the left leg kicking backward and the right leg forward. The half-twist ends with the trunk in a prone position.

6. *The Clearance, the Push-away and the Landing* (Fig. 96F and G). The clearance of the bar immediately follows the completion of the swing-up, the heave and the trunk-twist. Muscular effort rapidly diminishes as height is attained. The line of body progression in space is now parallel to that of the run-up and it is probable that the momentum of the run-up has much to do with the carriage of the vaulter's body over and across the bar. The push-away is not a part of the muscular effort of the vault. It is a necessary act to prevent the released pole from spoiling the event by hitting the bar. The landing is a not inconsiderable fall and the athlete must learn to accomplish it with the least danger to himself. The method of pole-vault, herein described as compared with the side swing vault, lessens this hazard.

This analysis of pole-vaulting must surely emphasize the statement of Conger[1]—namely: the details of the action (pole-vaulting)

[1] Roy M. Conger, M. A., Dir. of Recreational Sports, Penn. State College. A. S. Barnes & Co., N. Y., 1938.

consist in "mastering several techniques of the event and then molding them into one harmonious, syncronized whole". Probably no other athletic event requires more trained co-ordination of the entire body mechanism than does this one.

In its *relationship to body poise*, pole-vaulting may be classed in the second group of this book (see p. 241). The use of the pectoral muscles offsets the employment of the shoulder retracting muscles; the back-muscle activity is balanced by that of the abdomen, and the arm-muscle activities are about equal on both sides; but the shoulder-blade-controlling muscles and the twisting abdominal muscles are definitely used asymmetrically; and so, to some extent, must be considered as a hazard to the formation of left lateral curvature of the spine.

Although this *tendency* to spinal curvature is but *slight*, the vaulter could *neutralize* it by canoeing, using the right paddle more than the left, by calesthenics or by daily practicing the writer's "key-note" corrective exercise (see p. 228).

OTHER SPORTS PRESENTING GREATER TENDENCY TO HORIZONTAL DISPROPORTION AND LATERAL ASYMMETRY

(Comments, but without analysis)

RUNNING HIGH JUMP, in relation to good posture, depends upon which form of leg-swing and bar-clearance is used. If the athlete clears the bar by the use of the "scissors" jump, he runs up from the side, springs off from one foot and carries first one and then the other leg across the bar. The activity is definitely one-sided in the legs. If the "roll" is used, complicated movements of legs, of trunk-twist and of arms are employed. The action is made up of distinctly unilateral movements.

HURDLING, whether in the "high hurdles" or the "low hurdles", is definitely unilateral in both the leg mechanism and the body-twisting and arm activities, as practically all hurdlers train for the take-off leap from one foot rather than alternating. The high hurdles involve use of the body-heave forward and both arms are forward; in the low hurdles, one arm and the opposite leg are forward and the other arm and leg opposite to it are backward at the moment of crossing the hurdles. There is but little horizontal disproportion in muscle development.

PUTTING THE SHOT requires more muscle activity in the right leg and the right arm and shoulder-girdle, and emphasizes extreme effort in the trunk-twisting muscles. It is decidedly a unilateral muscle developer and, as such should be counteracted by symmetrical body building work.

THROWING THE DISCUS, although involving a different body mechanism, is dependent upon almost the same asymmetrical body development as does Putting the Shot. As such, it is classified as a bi-lateral exercise, which should be met by appropriate symmetrical activities.

JAVELIN THROWING adds a run-up to the asymmetries of the Shot Put and the Discus Throw. It is to be classed, in its effect upon body poise as were the two just mentioned (see above). (Note: The not infrequent "wrench" to the mechanism of the shoulder-girdle and the shoulder-joint, of the side holding the javelin, brings this event into increasing disfavor with the American Amateur Athletic Association.)

ARCHERY, historically much used as a sport and as a means of warfare, is nevertheless a one-sided activity, especially in the use of the shoulder-girdle and in the arms. If archers would practice opposite stances and opposite use of the shoulder-girdle and arm mechanisms, the sport would take a high place in the formation of good posture.

In Part III, some leading sports and athletic events have been presented, with short historical sketch, with kinesiological analysis, with notation of tendencies toward faulty posture and with suggestions as to remedy. Other sports have been named, with only brief comment on their relationship to body poise. All have been classified in this posture relationship. It is to be noted that, although sports and games cannot take the place of systematic calesthenics and drills, in all-round body development, sports and games are useful, that many of them have comparatively little harmful effect on body poise and that those which have can and should be associated with the use of counteracting physical development work.

GLOSSARY OF ANATOMICAL TERMS

Abduction: Movement of a part of the body, in the lateral plane, increasing the angle between the moving part and the axis of the body. It is the opposite of adduction.

Abductor hallucis (of the great toe): Muscle in base of foot, extends from heel bone (os calcis) to the bone of the great toe; spreads and flexes the great toe.

Abductors: Muscles which lead a part away from the axis of the body.

Acetabulum: The cup-shaped socket for the hip joint, made up of parts of the three segments (ilium, ischium and pubis) of the hip bone (os innominatum).

Acromion process: Outward extension of the shoulder blade (scapula), forming the point of the shoulder and giving attachment to muscles.

Adduction: Movement of a part of the body, in the lateral plane, lessening the angle between the moving part and the axis of the body. It is the opposite of abduction.

Adductor hallucis (of the great toe): Muscle of the base of the foot, extends from the mid forefoot (metatarsal bones) to the base of the great toe; draws the great toe center-ward.

Adductors: Muscles which bring a part toward the axis of the body.

Adductors of the thigh: Three muscles, Adductor magnus, A. longus and A. brevis, extending from the pelvis down the inner side of the thigh to the thigh bone (femur); draw the thigh inward.

Anatomy: Descriptive account of an organic body.

Anconeus: Muscle, extending from the lower end of the arm bone (external condyle of humerus), to the upper end of ulna (olecranon process); extends the forearm.

Articular processes: Two pairs, an upper and a lower, of bony outgrowths, found at the sides of the bony ring (neural arch) of each vertebra. They regulate the direction and the amount of movement between consecutive vertebrae.

Astragalus: Highest bone or "keystone" of the foot (tarsus), directly supporting the bones of the leg and the entire body superstructure, in standing positions.

Atlas: The highest vertebra of the spine; rests upon the axis and, upon it, rests the head (occipital bone).

Axis: Second neck (cervical) vertebra of the spine, situated between the atlas and the third cervical vertebra. Important bone, regulating head-turnings to the right and to the left.

Biceps (Biceps cubiti): "Two-headed" muscle at front of upper arm, connecting shoulder-blade (scapula) with outer bone of forearm (radius); flexes and outward turns forearm.

Biceps femoris: "Two-headed" muscle at the outer rear of thigh, connecting pelvis (ischium) with outer bone of leg (fibula); flexes and turns leg outward. One of the Hamstring muscles.

Brachialis anticus: Muscle, extending from the lower half of the arm bone (humerus), downward, in front of the elbow joint, to near the upper end of the ulna: bends (flexes) the forearm.

Calcaneo-astragaloid joint: Found between the heel bone (os calcis) and the keystone bone (astragalus) of the foot.

Calcaneo-cuboid joint: Found between the heel bone (os calcis) and the cuboid bone of the foot.

Calcaneum—see Os calcis.

Centrum: The body or supporting part of any vertebra; is separated from its neighbor, above and below, by elastic cartilage (intervertebral disc).

Cervical spine—Bones of the neck: That portion of the spine found between the head and the bones of the chest region (thoracic spine). Seven vertebrae.

Cervicales ascendens—Ascending muscle of the neck: Found at the rear of the neck, extending from the upper ribs, upward, backward and inward to the transverse processes of the middle cervical vertebra; extends the neck and head.

Circumduction: Cone-like circular movement of a body part (as the arm) about a more or less fixed point, which is the apex of the cone.

Clavicle (Collar bone): Slightly S-shaped bone at the upper front of the chest, united to the sternum at the sterno-clavicular joint, and to the shoulder blade at the acromio-clavicular joint. One of the two bones of the shoulder girdle.

Coccyx: Small bone at extreme lower end of spine.

Complexus: Muscle of the back of the neck, extending from the transverse processes of the upper thoracic and lower cervical vertebrae, upward and outward to the rear of the head (between the superior and inferior curved lines of the occiput); extends the head and twists to the same side.

Concentric contraction of muscle: Muscle active and shortening.

Coraco-brachialis: Muscle connecting shoulder-blade (coracoid process) with upper arm bone (brachium); draws upper arm forward and inward.

Coracoid process: Bony projection at the upper outer corner of the shoulder-blade (scapula), protecting the shoulder joint and affording attachment to muscles.

Coronoid process: Bony extension at the upper front of the ulna, which limits extreme flexion at the elbow joint.

Crureus muscle—see Quadriceps extensor.

Cuboid: The outer bone of the mid foot (tarsus).

Cuneiform: Three bones of the mid foot (tarsus).

Deltoid: Shoulder-cap muscle, connecting collar bone (clavicle) and shoulder-blade (scapula) with upper arm bone (humerus); lifts upper arm bone away from the side of the body (abducts the humerus).

Diaphragm: Large dome-like muscle found between the thoracic and the abdominal cavities, extending from an extensive base along the lower borders of the ribs and from the upper lumbar spine, upward and inward to a central tendon. It is a muscle primarily of breathing. When in action the central tendon is lowered, increasing the vertical diameter of the chest cavity, and is thus an important part of the intake of the breath (inspiration). It separates the abdominal and the chest cavities.

Dorsi-flexors: Muscles which lessen the angle between the front of the leg and the foot.

Disc, intervertebral—see Intervertebral disc.

Eccentric contraction of muscle: Muscle active and lengthening.

Erector spinae (Straighteners of the spine): Large group of muscles of the back, extending from the pelvis and lower spine, upward and dividing into two lesser muscle groups, to act upon rear of the mid spine and of the six lower ribs; straightens spine and bends spine backward (extension).

Extensor carpi radialis brevior: Muscle, extending from the lower end of the arm bone (external condyle of humerus), back of the wrist joint, to the hand (base of third metacarpal bone); extends the wrist and flexes the forearm.

Extensor carpi radialis longior: Muscle, extending from the lower end of the arm bone (external condyle of humerus), back of the wrist, to the index finger; extends the wrist and flexes the forearm.

Extensor carpi ulnaris: Muscle, extending from the lower end of the arm bone (external condyle of humerus), to the outer rear of the hand (fifth metacarpal bone); extends and abducts the wrist and flexes the forearm.

Extensor communis digitorum: Muscle, extending from the lower end of the arm bone (outer condyle of humerus), back of the wrist joint, to the second and third fingers; extends fingers and forearm.

Extensor indicis: Muscle, extending from the rear of the ulna, to the index finger; extends the index finger.

Extensor longus digitorum: Muscle extending from outer rear of leg bones (tibia and fibula) to the four lesser toes; extends toes, flexes and turns the foot outward.

Extensor minimi digiti: Muscle, extending from the lower end of arm bone (external condyle of humerus), back of the wrist, to the base of the little finger; extends the little finger.

Extensor ossis metacarpi pollicis: Muscle, extending from the back of the hand (third metacarpal bone), to the thumb; abducts the thumb.

Extensor primi internodii pollicis: Muscle, extending from the back of the radius, to the thumb; extends and abducts the thumb.

Extensor proprius hallucis: Muscle, extending from the front of the larger leg bone (tibia) to the base of the great toe; extends the great toe.

Extensor secundi internodii pollicis: Muscle, extending from the rear of the forearm to the thumb; extends the thumb.

Extension: Movement of a part of the body, in the front-back plane, increasing the angle between the moving part and the more fixed neighboring part. It is the opposite of flexion.

Extensors: Muscles which increase the angle, in the front-back plane of the body, between two or more connecting bones.

External oblique (Obliquus externus): Muscle of the abdomen, found just beneath the skin or fatty tissue, connecting lower ribs with the pelvis; compresses abdominal contents and bends and twists upper trunk. (The twist is in the direction opposite to that of the Internal oblique.)

Femur (Thigh bone): Longest bone of the body, extending from the hip joint to the knee joint.

Fibula: Smaller of the two bones of the leg, extending from just below the knee to the key-bone of the foot (astragalus), on the outer side of the leg.

Flexion: Movement of a part of the body, in the front-back plane, lessening the angle between the moving part and the more fixed neighboring part. It is the opposite of extension.

Flexor accessorius pedis: Same as Flexor brevis digitorum. Muscle of the sole of the foot, extending from the heel bone (os calcis) to the bases of the four outer toes; flexes the toes.

Flexor brevis hallucis: Muscle of the sole of the foot, extending from the inner mid foot (cuboid and external cuneiform) to the base of the great toe; flexes the great toe.

Flexor brevis minimi digiti: Muscle of the sole of the foot, extending from outer sole (fifth metatarsal) to the base of the little toe: flexes the little toe.

Flexor brevis pollicis: Muscle, extending from the front of the wrist to the thumb; flexes the thumb.

Flexor carpi radialis: Muscle, extending from the lower end of the arm bone (internal condyle of humerus), in front of the wrist joint, to the hand (second and third metacarpal bones); flexes and abducts the wrist, flexes and turns the forearm inward.

Flexor carpi ulnaris: Muscle, extending from the lower end of the arm bone (internal condyle of humerus), and upper end of ulna (olecranon process), in front of the wrist joint, to the wrist (fifth metacarpal and pisiform bones); flexes and adducts the wrist.

Flexor longus digitorum: Muscle of the inner rear of the leg, extending from the shin bone (tibia) downward at the rear of the inner ankle bone (internal malleolus) to the under surface of the four lesser toes; bends (flexes) the toes and turns the foot inward (supination).

Flexor longus hallucis: Muscle of the inner rear of the leg, extending from the lesser leg bone (fibula) downward at the rear of the inner ankle bone (internal malleolus) to the under surface of the great toe; bends (flexes) the great toe and turns the foot inward (supination).

Flexor longus pollicis: Muscle, extending from the radius and ulna (coronoid process), across the outer front of the wrist to the thumb; flexes the thumb.

Flexor profundus digitorum: Muscle, extending from the shaft of the ulna, across the front of the wrist, to the fingers; flexes the fingers.

Flexors: Muscles which lessen the angle, in the front-back plane of the body, between two or more connecting bones.

Flexor sublimis digitorum: Muscle, extending from the lower end of the arm bone (inner condyle of humerus) and from the middle of the radius, across the front of the wrist, to the fingers; flexes the fingers and the forearm.

Foramen magnum: Large opening at the base of the skull (occiput) through which passes the spinal cord.

Gastrocnemius (Calf muscle): The principal muscle of the calf of the leg, connecting the thigh bone with the heel of the foot. Its lower portion is the strong tendon of Achilles. With the Soleus, it increases the angle between the front of the leg and the foot (plantar flexes the foot).

Gemelli: Two muscles, Gemellus superior and G. inferior, extending from the middle of the outer surface of the pelvis, outward, above the hip joint, to the upper end of the thigh bone (great trochanter of femur); outward twists and separates the thigh bone (outward rotation and abduction).

Glutei: Three muscles, Gluteus maximus, G. medius and G. minimus, which constitute the mass of the buttock, extending between the pelvis and the thigh bone (femur); carry the thigh bone backward and outward. G. maximus is the principal outward rotator of the thigh.

Gracilis: Longest muscle of the body, extending from the pelvis (pubes) to the larger leg bone (tibia); carries the thigh bone (femur) forward outward (flexion and abduction).

Hamstring muscles: Three muscles at the back of the thigh, Biceps femoris, Semitendinosis and Semimembranosis, extending between the pelvis and the upper leg bones (tibia and fibu'a); flex (bend) the leg on the thigh.

Human kinesiology—see Kinesiology.

Humerus: Upper arm bone, connecting the shoulder-girdle (scapula), at the shoulder-joint, with the ulna and radius, at the elbow joint.

Hyoid bone: Small horseshoe-shaped bone in the front of the neck, which acts as a deflector of the line of pull of some of the neck muscles.

Iliacus: Muscle, extending from the broad inner surface of the pelvis (ilium), forward, uniting with the Psoas magnus, to the upper end of the thigh bone (lesser trochanter of femur); flexes and outward turns the thigh bone.

Ilio-psoas: Two muscles, Psoas magnus and Iliacus, having combined action, which extend from the lower spine (Psoas magnus) and the pelvis (Iliacus) to the thigh bone (femur); flex and turn the thigh bone outward (outward rotation).

Infracostales: Muscles found at the upper inner surface of the chest, extending from rib to rib; aid in raising the ribs in the act of breath intake (inspiration).

Infraspinatus: Muscle, extending from the shoulder-blade (infraspinous fossa), outward to the upper end of the arm bone (greater tuberosity of humerus); turns the arm outward (outward rotation).

In-rotators: Muscles which turn a bone or bones inward.

Intercostal muscles: Two sets of muscles, External and Internal, which pass from one rib to its neighbor and, by lifting (elevation) and turning the ribs outward (eversion), become muscles of breathing.

Internal oblique (Obliquus internus): Muscle of the abdomen, found beneath the External oblique, connecting lower ribs with pelvis; compresses abdominal contents and bends and twists upper trunk. (The twist is in the direction opposite to that of the External oblique.)

Interossei: Muscles of the base of the foot, extending from the mid foot (metatarsal bones) to the neighboring toes on either side; separate (adduct) the three outer toes. There are Interossei of the hand, with similar position and similar function to the fingers.

Intervertebral disc: Elastic cartilage found between consecutive vertebrae (centrum) to ease the effects of jarring and to allow pliability between neighboring vertebrae.

Kinesiology: The science of bodies in motion. Human kinesiology, the anatomical analysis of human positions and movements.

Lamina: That portion of a vertebra which connects the front portion of the bony ring (pedicle) with the spine (spinous process).

Latissimus dorsi (Broadest of the back): Muscle which connects the mid and lower spine, the rear lower ribs and the pelvis with the upper arm bone; draws the arm backward and downward.

Levator anguli scapulae (Elevator of the angle of the shoulder-blade): Muscle, extending from the transverse processes of the four upper neck (cervical) vertebrae, downward and outward to the posterior superior angle of the shoulder-blade (scapula); raises the upper angle of the shoulder-blade or twists the head to the opposite side.

Levatores costarum (Raisers of the ribs, twelve pairs): Muscles of the back of the neck and upper thoracic spine, extending from the transverse processes of the seventh cervical and eleven upper thoracic vertebrae, outward and downward to the next rib below; lift the ribs in in-breathing (inspiration).

Ligamentum nuchae: A broad fibrous band at the back of the neck which extends from the tips of the spinous processes of the cervical vertebrae to the rear of the head (external occipital crest).

Longus colli (Long muscle of the neck): Muscle, extending from the transverse processes of the middle neck (cervical vertebra) upward, forward and inward to the base of the skull (occiput); bends the head forward and twists to the same side.

Lumbar spine: That portion of the spine found between the spine of the chest region (thoracic spine) and the pelvis. Five vertebrae.

Lumbricales: Muscles of the base of the foot, extending from the tendon of the long flexor of the toes (flexor longus digitorum) to the bases of the lesser toes; aids the long flexors and extensors of the toes.

Mandible: Horseshoe-shaped bone, constituting the lower jaw.

Medulla oblongata: Important thickening at the upper end of the spinal cord, connecting the brain (cerebellum) and the remainder of the spinal cord.

Meniscus: Two crescent-shaped cartilages, external m. and internal m., which are located around the outer margins of the knee joint, and aid in regulating the action of the joint.

Metacarpal: Adjective referring to the five parallel bones of the hand between the carpus (wrist) and the fingers.

Metacarpus: That portion of the hand which forms the palm. Five bones.

Metatarsal: Adjective referring to the five parallel bones of the foot between the foot (tarsus) and the toes.

Metatarsus: That portion of the foot found between the tarsus and the toes. Five bones.

Multifidus spinae: Muscle of the back, extending from the sacrum, the pelvis and the transverse processes of lumbar, thoracic and cervical spine, upward outward each to the lamina of a vertebra above; extends the spine and twists each vertebra outward in relation to the vertebra below.

Navicular (Scaphoid): One of the bones of the mid foot (tarsus).

Naviculo-cuboid joint: Found between the navicular and the cuboid bones of the foot.

Neural arch: The bony ring, found in each vertebra, at the rear of the body (centrum), which encircles and protects the spinal cord and nerve-trunk exits. It is made up of two pedicles, two laminae and seven processes—a pair of transverse processes, two pairs of articular processes and a single spinous process.

Obliquus capitis inferior (Lower oblique muscle of the head): Muscle, extending from the spinous process of the axis (second bone of the neck) upward and outward to the transverse process of the atlas (top bone of the neck); twists (rotates) the atlas and head to the same side.

Obliquus capitis superior (Upper oblique muscle of the head); Muscle, extending from the transverse process of the atlas (top neck bone), upward, rearward and inward, to the base of the skull (occipit); twists the head to the opposite side.

Obturator externus: Muscle, extending from the middle of the pelvis outward at the rear of the hip joint to the upper end of the thigh bone (femur); turns the thigh outward (outward rotation).

Obturator internus: Muscle, extending from the middle of the pelvis, back of the hip bone to the upper end of the thigh bone (femur); turns outward and abducts the thigh (outward rotation—abduction).

Occiput (Occipital bone): Flat bone at the base of the skull, which rests upon the spine (atlas) and has an important opening (foramen magnum) through which passes the spinal cord.

Odontoid process: Tooth-like upper projection of bone from the axis (second neck vertebra), making joint connection with the atlas (first neck vertebra) and regulating lateral twist (rotation) of the head.

Olecranon process: Upward extension at the rear of the ulna, which prevents over-extension of the elbow joint and gives attachment to muscles.
Opponens minimi digiti: Muscle, extending from the front of the wrist, to the hand (fifth metacarpal bone); flexes the fifth metacarpal bone.
Os calcis (Calcaneum): Heel bone; rearmost of the tarsal bones of the foot.
Os innominatum (The hip bone) (plural ossa innominata): One of four bones which make up the body pelvis. It is divided into ilium and ischium and pubes, all of which three divisions constitute a part of the hip-joint socket (acetabulum).
Palmaris longus: Muscle, extending from the lower end of the arm bone (inner condyle of the humerus), in front of the wrist, to the palm of the hand; flexes the elbow and the wrist.
Patella (Knee cap): Compact bone, situated at the front of the knee joint, over which it glides, in flexion and extension activities of the joint. It extends the action and force of the great front thigh muscle (Quadriceps extensor), across the knee, to the upper front of the larger leg bone (tibia).
Pectorals: Two muscles, Pectoralis major and P. minor, found at the outer front of the chest. P. major extends from the breast-bone (sternum) and collar-bone (clavicle) to the upper arm bone (humerus); draws arm forward, downward and inward. P. minor extends from upper ribs in front to shoulder-blade (coracoid process of scapula); lowers the point of the shoulder.
Pedicle: That portion of a vertebra which connects the body (centrum) with the rear portion of the bony ring (lamina).
Pelvis: The large bony mechanism at the lower end of the trunk. It consists of the hip bone (ilium), lowest portion (ischium), front arch (pubes) and two separate bones, (sacrum and coccyx).
Peronei: Three muscles Peroneus longus, P. brevis and P. tertius, found at the outer front of the leg, extend from the two leg bones (tibia and fibula) to outer mid foot; increase the front leg-foot angle (plantar flex) and turn the mid foot outward (pronation), when the leg is partly flexed.
Phalanges: Fourteen bones of the fingers or of the toes; singular, phalanx.
Plantar: Adjective referring to the sole of the foot.
Plantar flexors: Muscles which increase the angle, at the ankle joint, between the front of the leg and the foot.
Plantaris: Small accessory muscle, extending from the rear of the thigh bone (femur) to the heel bone (os calcis); plantar flexes the foot.
Popliteus: Muscle at the rear of the knee joint, extending from the outer, lower end of the thigh bone (external condyle of femur) to the upper rear of the larger leg bone (tibia); flexes and inward rotates the leg.
Pronation: Movement of a body part, as the hand and forearm and the foot, to a "face down" position.

Pronator quadratus: "Four-sided inward-turning" muscle, at the front of the lower forearm, connecting the outer forearm bone (radius) with the inner forearm bone (ulna); turns the forearm and hand inward.

Pronator radii tares: "Round inward-turning" muscle of the forearm, connecting upper arm bone (humerus) with the forearm (radius) and the wrist; turns the forearm and hand inward.

Psoas magnus: Muscle, extending from the front of the lower spine downward and forward across the sides of the pelvis, uniting with the iliacus, to the upper end of the thigh bone (lesser trochanter of femur); flexus and turns the thigh bone outward (outward rotation).

Pyriformis: Muscle, extending from the inner surface of the pelvis, through the sciatic notch, to the upper end of the thigh bone (femur); turns the thigh outward (external rotation).

Quadratus lumborum: Muscle at the side of the abdomen, connecting the pelvis and the lower ribs; compresses the abdominal contents and bends trunk sideways.

Quadratus extensor (femoris): Large muscle group in front of thigh (Rectus femoris, Vastus externus, Vastus internus and Crueus) extends from the pelvis (ilium) and front and sides of the thigh bone (femur) to the knee-cap (patella). A strong band (ligamentum patellae), passing from the knee-cap to the front of the leg (tibia), continues the action of this muscle group; straightens the leg (extension at the knee).

Radial flexion: Sideways movement, at the wrist, toward the radial (thumb) side.

Radius: Bone of the forearm, uniting (indirectly) with the upper arm bone (humerus), with the upper and lower ends of the ulna and with the bones of the wrist (carpus).

Rectus abdominis: Muscle of the abdomen, extending up and down, at the middle of the front of the abdomen, from the pelvis (pubes) to the cartilages of the fifth, sixth and seventh ribs in front; compresses the abdominal contents and flexes the spine by indirect acton on the ribs.

Rectus capitis anticus major et minor (The large and small straight front muscles of the head): Two muscles, situated at the upper front of the neck, extending from the transverse process of the third and sixth neck (cervical) vertebrae and from the lateral mass of the atlas upward, forward and inward to the base of the skull (occiput); twist the head to the same side and draw it backward on the neck.

Rectus capitis lateralis (Straight muscle at the side of head): Muscle, extending from the transverse process of the atlas (top bone of the neck), upward and outward, to the base of the skull (occiput); bends the head sideways.

Rectus femoris: One of the four divisions of the great front thigh muscle group (Quadriceps extensor), connects the pelvis with the knee-cap (patella). (See also Quadriceps extensor.)

Rhomboids: Two muscles Rhomboideus major and R. minor, connecting the lower neck and upper dorsal spines with the shoulder-blade (scapula); raise and draw the shoulder-blade backward.

Rotation: Movement of a body part about its own axis. Inward rotation and outward rotation.

Rotatores spinae: Muscles of the back, extending from the transverse processes of the thoracic spine, each to the lamina of the vertebra next above; extend the upper portion of the spine and turn the neck and head to the opposite direction.

Rotators: Muscles which turn a bone inward (in-rotation) or outward (out-rotation).

Sacrum: Large triangular-shaped bone, constituting the base of the spinal column, snuggly fitted between the hip bones at the sides, and connecting the lumbar spine and the coccyx, respectively above and below.

Sartorius: "Tailor muscle," because used by tailors in their crossed legged sitting position while at work; extends from the pelvis (anterior superior spine of ilium) to the large leg bone (tibia); flexes and crosses the legs.

Scaleni muscles: Three muscles, Scalenus anticus, S. medius and S. minimus, situated at the sides of the neck, extending from upper ribs and transverse processes of lower cervical vertebrae, outward and upward, to transverse processes of higher cervical vertebrae; bend the head and neck sideways.

Scaphoid—See navicular bone.

Scapula: Shoulder-blade, one of the two bones of the shoulder-girdle and found at the upper outer rear of the chest; joined to the collar-bone (clavicle) and to the upper arm bone (humerus).

Semimembranosis: One of the Hamstring muscles at the inner rear of the thigh, extending from the pelvis (tuberosity of the ischium), to the larger of the leg bones (tuberosity of tibia); flexes the leg and rotates inward.

Semispinalis colli: Muscle, extending from the transverse processes of four lower cervical and six upper thoracic vertebrae, upward and inward, to the spinous processes of the second to fifth neck (cervical) vertebrae; extends the neck and twists the head to the opposite side.

Semispinalis dorsi: Muscle of the back, extending from the transverse processes of the lower thoracic vertebrae, upward and inward, to the spinous processes of the upper thoracic and lower cervical vertebrae; extends the spine and rotates the vertebra to the opposite direction.

Semitendinosis: One of the Hamstring muscles at the inner rear of the thigh, extending from the pelvis (tuberosity of ischium) to the larger of the leg bones (upper inner tibia); flexes leg on thigh.

Serratus magnus: "Saw-shaped" muscle found at the side of the mid chest, extending from eight upper ribs to the shoulder-blade (scapula); turns the shoulder-blade upward and, in reverse action, elevates ribs in in-breathing (inspiration).

Seratus posticus inferior: Muscles of the mid back, extending from the spines of two lower thoracic and three lumbar vertebrae, upward and outward, to the four lower ribs; lower the ribs in out-breathing (expiration).

Serratus posticus superior: Muscles of the upper back, extending from the spines of the seventh cervical and three thoracic vertebrae, downward and outward, to the rear of the second, third and fourth ribs; lift the ribs in in-breathing (inspiration).

Shoulder-girdle: Collar-bone (clavicle) and shoulder-blade (scapula). (Term used as of one side or as of both sides.)

Soleus: Smaller muscle of the calf of the leg; connects the leg bones (tibia and fibula) with the heel bone (os calcis). With the Gastrocnemius, it increases the angle, at the ankle joint, between the front of the leg and the foot (plantar flexes the foot).

Spinalis dorsi: Muscle of the back, extending from the upper lumbar and lower thoracic vertebrae (spinous processes), upward to the remaining thoracic spinous processes; extends the spine.

Spinous process: Projection of each vertebra which extends backward and can usually be felt below the skin. Its function is protection and to afford leverage for muscle attachments.

Splenius capitus et colli: Combined muscles of the rear of the neck and upper thoracic vertebrae, extending from the lower half of the ligamentum nuchae, spines of the seventh cervical and first three thoracic vertebrae, upward outward, to the rear and sides of the skull (middle oblique line and mastoid process); extend the head and neck and rotate the head to the same side.

Sterno-hyoid: Small muscle at the front of the neck, extending from the breast-bone (sternum) and collar-bone (clavicle), upward to the hyoid bone; flexes the head and neck and depresses the larynx.

Sterno-mastoid: Muscle at front of neck, extending from the breast-bone (sternum) and collar-bone (clavicle), upward and outward, to the side of the skull (mastoid process of temporal bone); flexes the head and neck and rotates to the opposite side.

Sternum (Breast-bone): Flat bone at the front of the chest to which many of the ribs are united by elastic cartilage.

Subclavius: Small muscle, extending from the first rib, upward and backward, to the lower surface of the collar-bone (clavicle); draws the collar-bone downward and forward.

Subscapularis: Muscle, extending from the shoulder-blade (scapula) to the upper end of the arm bone (lesser tuberosity of humerus); rotates the upper arm bone inward and draws the arm to the side of the body (adduction).

Supination: Movement of a part of the body, as the hand and forearm and the foot, on its own axis, to an "on the back" position.

Supinators of the forearm: Two muscles, Supinator longus and S. brevis, found at the outer rear of forearm, connecting the upper arm bone (humerus) with the outer forearm bone (radius) and the wrist; flex and turn the forearm outward.

Supraspinatus: Muscle, extending from the shoulder-blade (superspinous fossa of scapula) outward to the arm bone (greater tuberosity of the humerus); draws the arm away from the body (abduction of shoulder joint).

Sustentaculum tali: Front portion of the heel bone (os calcis).

Tarsal: Adjective referring to the tarsal bones of the foot.

Tarsus: Seven principal bones of the foot.

Temporal bones: Flat and irregular-shaped bones on each side of the skull, affording attachments to muscles which move the head and supporting the ears and important in the internal ear mechanism.

Tendo Achilles: Strong downward tapering band, at the lower rear of the leg, extending the action of the Calf muscles (Gastrocnemius and Soleus), to the heel bone (os calcis); plantar flexes the foot, at the ankle joint.

Tensor vaginae femoris (Tensor fasciae latae): Muscle on the outer side of the thigh, extending from the pelvis (anterior superior spine of ilium) to the strong ligamentous band (fascia lata), which encloses the large muscles of the outer side of the thigh; tightens this ligamentous band and tends to carry the thigh bone outward (abduction).

Teres major: Muscle, extending from the shoulder-blade (scapula), upward, outward and forward, to the shaft of the upper arm bone (humerus); draws the arm downward backward (extension) and rotates it inward.

Teres minor: Muscle, extending from the shoulder blade (scapula), to the upper end of the upper arm bone (greater tuberosity of humerus); rotates outward and draws the arm away from the side of the body (abduction).

Thoracic spine: That portion of the spine found between the bones of the neck (cervical spine) and the bones of the loin (lumbar spine). It is a part of the chest (bony thorax).

Thorax (Bony thorax): The chest or bony cage made up of twelve thoracic vertebrae, twelve ribs and sternum. It contains the lungs, the heart and large blood vessels. At its base is the dome-shaped muscle and tendon, known as the Diaphragm, which separates its contents from those of the abdomen.

Tibia (Shin bone): Larger of two bones of the leg, extending from the thigh bone (femur) to the key-bone of the foot (astragalus). Inner side of the leg.

Tibialis anticus: Muscle of the inner front of the leg, extending from the shin bone (tibia) to the inner and front inner mid foot; lessens the front leg-foot angle, at the ankle joint, (dorsi flexes) and turns the mid foot inward (supination).

Tibialis posticus: Muscle of the inner rear of the leg, extending between both bones of the leg (tibia and fibula) to the inner rear of the mid foot; increases the front leg-foot angle, at the ankle joint, and turns the rear of the foot inward (supination).

Tonic contraction of muscle: Muscle active, but without changing length.

Trabiculae: Fine lines, in the internal structure of bones, to combine strength or resistance with lightness.

Trachelo-mastoid: Muscle of the back of the neck, extending from transverse processes of the upper thoracic and lower vertebral spines, upward and outward, to the side of the head (mastoid process); extends the head and neck and twists to the same side.

Transversalis abdominis: Muscle of the abdomen, found between the Internal oblique muscle and the abdominal peritoneum, extending from the lower ribs, the lumbar spine and the pelvis, inward, to unite with its fellow of the opposite side; compresses the abdominal contents.

Transverse processes: A pair of bony projections extending sideways from the bony ring (neural arch) of each vertebra. Their function is protection and to afford leverage for muscle attachments.

Transversus pedis: Muscle, extending across the bases of the toes, from the plantar ligament to the base of the great toe; adducts (draws centerward) the great toe.

Trapezius: Large fan-shaped muscle, at the rear of the upper trunk and the neck, connecting the upper spine and the head with the shoulder-blade (scapula) and the collar-bone (clavicle); draws the shoulder-blade backward and turns it upward; also, in reverse action, the upper portion lifts a droop-head (extends the head and neck).

Triangularis sterni: Small muscles at the front of and within the chest, extending from the lower end of the breast-bone (sternum) to the cartilages of the second to the sixth ribs; narrow the chest in out-breathing (expiration).

Triceps: "Three-headed" muscle at the rear of the upper arm, connecting the shoulder-blade (scapula) and upper arm bone (humerus) with the inner forearm bone (ulna); straightens (extends) the forearm.

Ulna: Bone of the forearm, uniting with the humerus, at the elbow joint, with the radius, at the sides of the upper and lower ends, and with the bones of the wrist (carpus).

Ulnar flexion: Sideways movement, at the wrist, toward the ulnar (little finger) side.

Vastus externus muscle—see Quadriceps extensor.

Vastus internus muscle—see Quadriceps extensor.

Vertebra: Any single bone of the spine; plural, vertebrae.

Vertebral column: The spine as a whole.

INDEX

Abductor hallucis muscle, 18
Abductor pollicis muscle, 108, 110
Acetabulum, 38
Acromio-clavicular joint, 92, 93, 94, 95
Acromio-coracoid ligament, 89
Acromion process, 86, 87, 89, 90
Adductor brevis muscle, 27, 36, 39, 40
Adductor hallucis muscle, 18
Adductor longus muscle, 27, 36, 39, 40
Adductor magnus muscle, 27, 36, 39, 40
Adductor minimi digiti muscle, 18
Adductor pollicis muscle, 108, 110, 111
Air hunger, 66
Allen, Ethan, 275
Anatomist's snuff box, 108
Anconeus muscle, 91, 99, 102
Ankle joint, 22, 23, 256
  Defined, 22
  Leverage of, 23
  Ligaments of, 22
  Kinesiology of, 23
  Muscles of, 22
Antaeus, 101
  kinesiology of, 123
Anterior arch of the atlas, 75, 80
Anterior atlanto-axoidean ligament, 75, 81
Anterior chondro-sternal ligament, 61, 65
Anterior common ligament, 36, 47, 48, 51, 54, 75, 80
Anterior costo-vertebral ligament, 61
Anterior crucial ligament of the knee, 30
Anterior ligament
  of ankle joint, 22
  of elbow, 101
Anterior oblique ligament, 76, 79
  of the neck, 81
Anterior occipito-atlantol ligament, 76, 79, 81
Anterior radio-carpal ligament, 100
Anterior radio-ulnar ligament, 100
Anterior transverse ligament, 17
Apparatus for foot measurement, 143
Archery, 241, 285

Arches of foot
  metatarsal, 17
  mid-transverse, 16
  plantar, 15
Arm, bones of, 85
Arm pit, 89
Articular fibro-cartilage, 90, 92, 95
Articular process
  of cervical spine, 75
  of lumbar spine, 45
  of thoracic spine, 53
Astragalus, 22
Asymmetrical disproportion of muscular development, 240, 261
Atlanto-axoidean joint, 80
Atlanto-odontoid ligament, 75
Atlas, 74, 76, 81
Auditory meatus, 78
Axilla, 89
Axis, 74, 75, 76, 83

Baseball, 240, 241, 273
  kinesiology of pitching, 274
Base of support, 9
Basi-occipital process, 78
Basketball, 241, 253
Ben Hur, 255
Biceps muscle, 87, 89, 96, 97, 100, 101, 104, 117, 247, 248, 251, 259, 265, 267, 271, 282, 283
  compound action of, 117
Biceps femoris muscle, 21, 27, 30, 36, 39, 40, 42
Bicycling, 240, 241, 272
Bodies of vertebrae
  of Cervical spine, 74
  of Lumbar spine, 45
Bony thorax, 60
  ligaments of, 61
Boxing, 240, 241, 268
  kinesiology of, 269
Braces and corsets, in treatment of scoliosis, 210
Brachialis anticus muscle, 91, 99, 101

INDEX

Brachio-radialis muscle, 101
Breast Bone, 61
Breast stroke, 249
　kinesiology of, 249

Canoeing, 240, 241, 254
Capitellum of humerus, 100
Capsular ligament, 27, 47, 48, 51, 54, 61,
　　64, 75, 76, 79, 80, 81, 87, 90, 92, 93,
　　94, 95, 96, 99, 105
　of the knee, 29
　of the hip joint, 36
Carpo-metacarpal joint, 107
　movements of, 108
　muscles of, 108
Carpus bone, 104, 105, 106
Cartilage
　articular fibro, 90, 92, 95
　cotyloid, 36, 38
　fibro, 55, 64, 74, 75, 80
　glenoid fibro, 89
　interarticular, 47, 51, 54, 61, 75, 87, 92, 93
　intervertebral fibro, 36, 47, 48, 53, 54, 75
　intervertebral fibro, of pelvis, 36
　semi-lunar of knee, 29
Central odontoid ligament, 76, 79, 81
Centrum of cervical vertebrae, 74
Cervical joints, 80
Cervical spine, 74
　muscles of, 76
　bones of, 85
Cervical vertebrae, 74
　articular surfaces of, 75
　body of, 74
　centrum, 74
　lamina of, 75
　neural arch of, 74
　pedicle of, 75
　spinous process of, 75
　transverse process of, 75
Cervicales ascendens muscle, 62, 76
Cervico-thoracic joint, 76
Check ligament, 76, 79, 81
Chest, 60, 258
　bones of, 85, 93
Clavicle bone, 86, 87, 92, 93, 258
Coccyx, 37

Complexus muscle, 54, 76, 79, 82, 83, 84
Compound action of muscles, 114
　biceps, 117
　forearm, wrist and hand, 119
　latissimus dorsi, 114
　pectoralis major, 116
　trapezius, 116
　triceps, 118
Conger, Roy M., 283
Conoid ligament, 87
Coraco-brachialis muscle, 89, 91, 96, 97, 251, 267, 271, 283
Coraco-humeral ligament, 89, 96
Coracoid process, 89, 96
Coronoid fossa, 101
Coronoid process, 79, 99, 101
Corrective gymnastics, 112
　for feet, 157, 158, 159, 160, 161, 162
Costo-central joint, 63, 64
　movements of, 64
Costo-clavicular ligament, 87, 92
Costo-sternal joint, 6, 63
Costo-transverse joint, 63, 64
　movements of, 64
Cotyloid cartilages, 36, 38
Crawl stroke, 242
Crew rowing, 255
　kinesiology of, 256
Crucial ligament, 76, 79
　of neck, 81
Crureus muscle, 27
Cubit, 92
Cuneiform bone, 104
Curves of Lumbar spine, 46
Cycling, 240, 241, 272

Debutante slouch, 32
Deltoid ligament, 22
Deltoid muscle, 87, 91, 96, 97, 98, 247, 248, 249, 267, 271, 276, 277
Diamond foot, 141
Diaphragm, 61, 62, 63, 67, 68
　depression of, 67
Disproportion of muscular development
　asymmetrical, 240, 261
　horizontal, 240, 261
　lateral, 240, 261
Distal and proximal, named, 109

Diving, 253
Doggett, Thomas, 255

Elbow joint, 85, 101, 258, 259
  ligaments of, 101
  movements of, 101
  muscles of, 101
Elkington's spinometer, 173
Erector spinal muscle, 36, 48, 49, 51, 54, 56, 57, 58, 62, 76, 82, 169, 258, 264, 274, 277
Exercises,
  de-rotation, 216, 218, 221, 223, 225, 226, 227, 228
  for faulty posture, 180, 181, 183, 184, 185, 186, 187, 190, 191
  for foot, 157, 158, 159, 160, 161, 162
  for posture training, 194, 195, 196
  for scoliosis, 211, 214, 215, 216, 218, 221, 223, 225, 226, 227, 228, 231, 233
Exercises as anatomical illustrations, arm raising sideways upward, analyzed, 112
Expiration, respiratory, 63, 69
Extensor brevis pollicis muscle, 107, 108, 110
Extensor carpi radialis brevior muscle, 91, 107
Extensor carpi radialis longior muscle, 91, 107
Extensor carpi ulnaris muscle, 91, 102, 105, 107
Extensor communis digitorum muscle, 91, 107, 109
Extensor indicis muscle, 99, 107, 109
Extensor longus digitorum muscle, 19, 21, 23
Extensor longus pollicis muscle, 104, 107, 108, 110
Extensor minimi digiti muscle, 91, 102, 107, 109
Extensor ossis metacarpii pollicis, 99, 100, 105, 107
Extensor primi internodii pollicis muscle, 100
Extensor proprius hallucis muscle, 19, 21
Extensor secundi internodii pollicis muscle, 99

External intercostal muscle, 62, 66
External lateral ligament
  of ankle joint, 22
  of wrist, 100
External oblique muscle, 247, 261, 265, 266, 276, 283

Faulty posture (see Posture, faulty)
Femur, 27, 38, 167, 256
  greater trochanter, 27
  lesser trochanter, 27
  ligaments of, 27
Fencing, 272
Fibro cartilage, 55, 64, 74, 75, 80
Fibula, 21, 256
Field hockey, 272
Fingers
  bones of, 85, 104, 107, 108
  joints of, 109
Flat foot, 138
  exercises, 157, 158, 159, 160, 161, 162
  hallux valgus, 141
  metatarsalgia, 138, 141
  metatarsus primus varus, 141
  padding and strapping, 148
  ratios, 142
  spastic, 163
  supports, 152
Flexor accessorius pedis muscle, 18
Flexor brevis digitorum muscle, 18
Flexor brevis hallucis muscle, 18
Flexor brevis minimi digiti muscle, 18, 105
Flexor brevis pollicis muscle, 105, 106, 108, 110
Flexor carpi radialis brevior muscle, 105
Flexor carpi radialis longior muscle, 91, 101, 105, 107
Flexor carpi ulnaris muscle, 91, 101, 105, 107
Flexor longus digitorum muscle, 18, 21, 22
Flexor longus hallucis muscle, 18, 21, 22
Flexor longus pollicis muscle, 99, 107, 108, 110
Flexor minimi digiti muscle, 109
Flexor profundus digitorum muscle, 99, 100, 107, 109

302 INDEX

Flexor sublimis digitorum muscle, 91, 99, 101, 107, 109
Foot
  apparatus for measurement of, 143
  arches of, 15
  bones of, 15, 85
  diamond, 141
  exercises, 157, 158, 159, 160, 161, 162
  flat, 138
  hallux valgus of, 141
  ligaments of, 17, 141
  movements of, 18
    abduction "pronation," 18
    adduction "supination," 18
    extension of mid-tarsus, 19
    extension of toes, 19
    flexion of mid-tarsus, 18
    flexion of toes, 18
  muscles of, 18, 140, 141, 142
  padding and strapping of, 148
  preventive measures for disability, 156
  ratios, 142
  shoes for, 152
  strains in relation to higher body strain, 155
  supports for, 152
  weak, 138
Football, 240, 241, 272
Foot disabilities, preventive measures, 156
Foot strains in relation to higher body strain, 155
Foot supports, 152
Foramen magnum, 78, 81
Forbes, A. MacKenzie
  explanation of mechanism of rotary lateral curvature of spine, 70
  history of mechanism of scoliosis, 211
Forearm
  bones of, 85, 198, 259
  compound action of muscles, 119
Fossa, 87
Frog kick in swimming, 251
Functional lateral curvature of spine, 167, 170
  c-curve, 171
  s-curve, 171
  summary of, 198

Fundamental gymnastic positions, 125
  fundamental half-kneeling, 128
  fundamental half-prone lying, 132
  fundamental half-standing, 125
  fundamental kneeling, 127
  fundamental lying, 130
  fundamental on hands and knees, 129
  fundamental prone-lying, 130
  fundamental sitting, 133
  fundamental spring sitting, 134
Fundamental positions
  gymnastic, 125, 127, 128, 129, 130, 132, 133, 134
  half-kneeling, 128
  half-prone lying, 9, 132
  half-standing, 125
  kneeling, 9, 127
  lying, 9, 130
  on hands and knees, 9, 129
  prone lying, 9, 130
  sitting, 9, 133
  spring sitting, 9, 134
  standing, 9, 13
Fundamental standing position
  bases of support, 13
  defined, 13
  line of gravity, 13

Gastrocnemius muscle, 22, 27, 30, 247, 265
Gemelli muscle, 27, 36, 39, 40
Gleno-humeral ligament, 89, 91
Glenoid cavity, 86, 95, 96
Glenoid fibro-cartilage, 89
Glenoid ligament, 106
Glenoid process, 89
Glutei muscles
  maximus, 27, 36, 39, 40, 243, 245, 248, 264, 265, 266, 274, 276, 277, 282
  medius, 27, 36, 39, 243, 245, 248, 264, 265, 266, 274, 276, 277, 282
  minimus, 27, 36, 39, 243, 245, 248, 264, 265, 266, 274, 276, 277, 282
Golf, 240, 241, 262
  kinesiology of the drive, 262
Gorilla type posture, 32, 167, 177
Gracilis muscle, 21, 30, 36, 39
Gravity
  center of, 9
  line of, 9

INDEX

Half-kneeling position, 128
Half-prone lying position, 9, 130
  defined, 10
  fundamental gymnastic, 132
Half-standing position, 125
  fundamental gymnastic, 125
Hallux valgus, 141
Hamstring muscle, 42, 247, 251
Hand
  bones of, 85, 104, 105, 259
  compound action of muscles, 119
Head, 77
  bones of, 85
  ligaments of, 79
  muscles of, 79
Head and neck
  kinesiology of, 84
  movements of, 81, 84
  muscles of, 81
Hercules, 121
  kinesiology of, 121
Hip joint, 38, 256
  kinesiology of, 42
  leverage of, 40
  ligaments of, 38
  movements of, 39
  muscles of, 39
Hockey, 272
  field, 272
  ice, 272
Horizontal disproportion of muscular development, 240, 261
Horse-back riding, 256
Humerus, bone, 86, 89, 92, 95, 96, 100, 101, 105, 107, 258, 259
  greater tuberosity of, 90, 96
  lesser tuberosity of, 90
  surgical neck of, 91
  fiberosities of, 90
Hurdling, 241, 284

Ice hockey, 272
Iliacus muscle, 36
Ilio-lumbar ligament, 36, 48
Ilio-psoas muscle, 27, 39, 170, 243, 244, 245, 266, 282
Inferior chondro-sternal ligament, 61, 65

Inferior radio-ulnar joint, 85, 98, 102, 103
Infracostales muscle, 62
Infraspinatus muscle, 87, 90, 97, 98
Infraspinous ligament, 36, 48, 54, 75, 80
Inspiration, respiratory, 62
Interarticular cartilage, 47, 51, 54, 61, 75, 87, 92, 93
Interclavicular ligament, 87, 92
Internal intercostal muscle, 62, 66
Internal lateral ligament
  of ankle joint, 22
  of elbow, 101
  of wrist, 99, 100
Internal oblique muscle, 247, 265, 266, 276, 283
Interossei muscle, 18, 106, 109
Interosseous ligament, 105
Intervertebral fibro-cartilage, 36, 47, 48, 53, 54, 75
Intervertebral fibro-cartilage of the pelvis, 36

Javelin throwing, 240, 241, 285
Joints
  acromio-clavicular, 92, 93, 94, 95
  ankle, 22, 256
  atlanto-axoidean, 80
  carpo-metacarpal, 107
  cervical, 80
  cervico-thoracic, 76
  costo-central, 63, 64
  costo-sternal, 63
  costo-transverse, 63, 64
  elbow, 85, 101, 258, 259
  fingers, 109
  hip, 38, 256
  inferior radio-ulnar, 85, 98, 102, 103
  knee, 29, 256
  lumbar spine, 50
  lumbo-sacral, 35, 258
  lumbo-thoracic, 55
  metacarpo-phalangeal, 85
  neck, 80
  occipito-atlantoid, 81
  occipito-odontoid, 81
  phalangeal, 85
  radio-ulnar, 102, 103
  sacro-lumbar, 48
  scapulo-clavicular, 92

Joints, shoulder, 85, 86, 92, 259
    sterno-clavicular, 85, 92, 94, 258
    superior radio-ulnar, 85, 98, 101, 102
    thoracic, 56
    thumb, 110
    wrist, 85, 99, 104, 106, 107
Joints of neck, 80
Jugular process, 78

Kangaroo type posture, 167, 177
Keynote exercise, author's, 229, 268
Kinesiology, 2
    corrective gymnastic, 9
    gymnastic, 9
    human, 9
Knee joint, 29, 256
    defined, 29
    kinesiology of, 32
    leverage of, 33
    ligaments of, 29
    movements of, 30
    muscles of, 29
Kneeling position, 9, 127
    defined, 10
    fundamental gymnastic, 127

Lamina
    of cervical spine, 75
    of lumbar spine, 45
    of thoracic spine, 53
Lateral asymmetry, 278
Lateral common ligament, 54, 75, 80
Lateral disproportion of muscular development, 240, 261
Lateral odontoid ligament, 76, 79, 81
Latissimus dorsi muscle, 36, 48, 49, 51, 54, 58, 62, 67, 72, 89, 91, 92, 93, 95, 96, 97, 98, 114, 247, 251, 258, 259, 261, 265, 266, 271, 276, 277, 282, 283
    compound action of, 114
Leg
    bones of, 21, 85
    defined, 19
Levator anguli scapulae muscle, 76, 82, 92, 95
Levator costarum muscle, 54, 62, 66
Leverage, 10
    moment of power, 12
    moment of weight, 12

Lever arms, 11
    of power, 11
    of weight, 11
Levers, 10
    orders defined, 10
Ligaments
    acromio-coracoid, 89
    anterior atlanto-axoidean, 75, 81
    anterior chondro-sternal, 61, 65
    anterior common, 36, 47, 48, 51, 54, 75, 80
    anterior costo-vertebral, 61
    anterior crucial of the knee, 30
    anterior occipito-atlantal, 76, 79, 81
    anterior oblique, 76, 79, 81
    anterior of ankle joint, 22
    anterior of elbow, 101
    anterior radio-corpal, 100
    anterior radio-ulnar, 100
    anterior transverse, 17
    atlanto-odontoid, 75
    capsular, 27, 47, 48, 51, 54, 61, 64, 75, 76, 79, 80, 81, 87, 90, 92, 93, 94, 95, 96, 99, 105
    capsular of hip joint, 36, 38
    capsular of knee, 29
    central odontoid, 76, 79, 81
    check, 76, 79, 81
    conoid, 87
    coraco-humeral, 89, 96
    costo-clavicular, 87–92
    crucial, 76, 79, 81
    deltoid of ankle joint, 22
    external lateral of ankle joint, 22
    external lateral of wrist, 100
    gleno-humeral, 89, 91
    glenoid, 106
    ilio-lumbar, 36, 48
    inferior chondro sternal, 61, 65
    infraspinous, 36, 48, 54, 75, 80
    internal lateral of ankle joint, 22
    internal lateral of elbow, 101
    internal lateral of wrist, 99, 100
    interossious, 105
    lateral, 76, 79, 81
    lateral common, 75, 80
    ligaments of shoulder joint, 96
    ligamentum subflava, 36, 48, 54, 75, 80
    ligamentum teres, 27, 36, 38

INDEX

long plantar, 17
medio-carpal posterior, 105
middle costo-transverse, 54, 61, 64
occipito-cervical, 76, 79, 81
orbicular, 99, 100, 101, 102
palmar, 105
plantar, 141
posterior atlanto-axoidean, 75, 81
posterior chondro-sternal, 61, 65
posterior common, 36, 47, 48, 51, 54 75, 80
posterior costo-transverse, 54, 61, 64
posterior crucial of knee, 30
posterior occipito-atlantol, 76, 79, 81
posterior of ankle joint, 22
posterior of elbow, 101
posterior radio-ulnar, 100
ponpart's, 36
rhomboid, 87, 92, 93, 97
sacro-sciatic, 36
sterno-clavicular, 87, 92, 93
superior chondro-sternal, 61, 65
superior costo-transverse, 54, 61, 64
supraspinous, 36, 48, 54, 75, 80
transverse, 27, 105
transverse of hip joint, 36, 38
transverse of neck, 75, 81
trapezoid, 87
Ligaments of shoulder joint, 96
Ligamentum subflava, 36, 48, 54, 75, 80
of the spine, 36
Ligamentum teres, 27, 36, 38
Long plantar ligament, 17
Longus colli muscle, 76, 82, 83
Lumbar spine, 45
articular process, 45
bodies of vertebrae, 45
bones of, 85
curves of, 46
lamina, 45
ligaments of, 46
neural arch, 45
pedicles, 45, 46
Lumbar spine joints, 50
joints of, 50
kinesiology of, 52
leverage of, 52
ligaments of, 51

movements of, 51
muscles of, 51
Lumbo-sacral joint, 35, 258
Lumbo-thoracic joint, 56
kinesiology of, 56
Lumbricales muscle, 18, 109
Lungs, 62
atmospheric pressure of, 62
Lying position, 9, 130
defined, 10
fundamental gymnastic, 130

Malleoli
external, 22
internal, 22
Mandible, 77, 78
Mastoid process, 78
Measurement of scoliosis, 199
x-ray examination, 201
Medio-carpal posterior ligament, 105
Medulla ablongata, 78, 81
Menisci, 29
Metacarpal bone, 104, 105, 106, 107
Metacarpo-phalangeal joint, 85
Metatarsal arch of foot, 17
Metatarsalgia, 138, 141
Metatarsus primus varus, 141
Middle costo-transverse ligament, 54, 61, 64
Mid-transverse arch of foot, 16
Moment of power, 12
Moment of weight, 12
Multifidus spinale muscle, 36, 54, 56, 57, 58, 59, 72, 73, 76, 82, 83
Muscle re-education in relation to protection, 6
Muscles, 36
origin and insertion defined, 36
Muscles
abductor hallucis, 18
abductor pollicis, 108, 110
adductor brevis, 27, 36, 39, 40
adductor hallucis, 18
adductor longus, 27, 36, 39, 40
adductor magnus, 27, 36, 39, 40
adductor minimi digiti, 18, 109
adductor pollicis, 108, 110, 111

INDEX

Muscles, anconeus, 91, 99, 102
  biceps, 87, 89, 96, 97, 100, 101, 104, 117, 247, 248, 251, 259, 265, 267, 271, 282, 283
  bicps femoris, 21, 27, 30, 36, 39, 40, 42
  brachiolis anticus, 91, 99, 101
  brachio-radialis, 101
  cervicales ascendens, 62, 76
  complexus, 54
  coraco-brachialis, 89, 91, 96, 97, 251, 267, 271, 283
  crureus, 27
  deltoid, 87, 91, 96, 97, 98, 247, 248, 249, 267, 271, 276, 277
  erector spinae, 36, 48, 49, 51, 54, 56, 57, 58, 62, 76, 82, 169, 258, 264, 274, 277
  extensor brevis pollicis, 107, 108, 110
  extensor carpi-radialis brevior, 91, 107
  extensor carpi-radialis longior, 91, 107
  extensor carpi ulnaris, 91, 102, 105, 107
  extensor communis digitorum, 91, 107, 109
  extensor indicis, 99, 107, 109
  extensor longus digitorum, 19, 21, 23
  extensor longus pollicis, 104, 107, 108, 110
  extensor minimi digiti, 91, 102, 107, 109
  extensor ossis metacarpii pollicis, 99, 100, 105, 107
  extensor primi internodii pollicis, 100
  extensor proprius hallucis, 19, 21
  extensor secundi internodii pollicis, 99
  external intercostal, 62, 66
  external oblique, 247, 261, 265, 266, 276, 283
  flexor accessorius pedis, 18
  flexor brevis digitorum, 18
  flexor brevis hallucis, 18
  flexor brevis minimi digiti, 18, 105
  flexor brevis pollicis, 105, 106, 108, 110
  flexor carpi-radialis brevior, 105
  flexor carpi-radialis longior, 91, 101, 105, 107
  flexor carpi-ulnaris, 91, 101, 105, 107
  flexor longus digitorum, 18, 21, 22
  flexor longus hallucis, 18, 21, 22
  flexor longus pollicis, 99, 107, 108, 110
  flexor profundus digitorum, 99, 100, 107, 109
  flexor sublimis digitorum, 91, 99, 101, 107, 109
  gastrocnemius, 22, 27, 30, 247, 265
  gemelli, 27, 36, 39, 40
  glutei, 27, 36, 39, 40, 243, 245, 247, 251, 258, 264, 265, 266, 272, 276, 277, 282
  gracilis, 21, 30, 36, 39
  hamstring, 42, 247, 251
  iliacus, 36
  ilio-psoas, 27, 39, 170, 243, 244, 245, 266, 282
  infracostales, 62
  infraspinatus, 87, 90, 97, 98
  internal intercostal, 62, 66
  internal oblique, 247, 265, 266, 276, 283
  interossei, 18, 106, 109
  latissimus dorsi, 36, 48, 49, 51, 54, 58, 62, 67, 72, 89, 91, 92, 93, 95, 96, 97, 98, 114, 247, 251, 258, 259, 261, 265, 266, 271, 276, 277, 282, 283
  levator anguli scapulae, 76, 82, 92, 95
  levator costarum, 54, 62, 66
  longus colli, 76, 82, 83
  lumbricales, 18, 109
  multifidus spinae, 36, 54, 56, 57, 58, 59, 72, 73, 76, 82, 83
  obliquus capitus inferior, 76, 83
  obliquus capitus superior, 76, 79, 84
  obliquus externus, 49, 51, 58, 59, 62
  obliquus internus, 36, 49, 51, 58, 59, 62, 73
  obturetor externus, 27, 36, 40, 57
  obturetor internus, 36, 49, 51, 58, 59, 62, 73
  opponens minimi digiti, 105, 106
  opponens pollicis, 105, 108, 110
  palmaris longus, 91, 101, 107
  pectineus, 36
  pectoralis major, 62, 67, 72, 87, 89, 91, 93, 94, 95, 96, 97, 98, 116, 247, 249, 251, 265, 266, 267, 271, 276, 277, 282, 283
  pectoralis minor, 67, 92, 93, 94, 247, 249, 276, 277, 282, 283

INDEX 307

peroneus brevis, 18, 19, 21, 23, 141
peroneus longus, 18, 19, 21, 23, 141
peroneus tertius, 18, 19, 21, 23, 141
plantaris, 27
popliteus, 21, 27
pronator quadratus, 99, 100, 103, 249, 251
pronator radii teres, 91, 99, 100, 101, 103, 249, 251
psoas magnus, 36, 42, 48, 49, 51
pyriformis, 27, 36, 40
quadratus lumborum, 36, 48, 49, 51, 57, 58, 59, 62, 274
quadriceps, 243, 244, 245, 246, 247, 252, 265, 266, 274, 276, 277, 282
quadriceps, femoris 21, 27, 30, 36, 39, 40
rectus abdominis, 49, 51, 57, 62
rectus capitus anticus major, 76, 79, 82, 83, 84
rectus capitus anticus minor, 76, 79
rectus capitus lateralis, 76, 79, 84
rectus femoris, 36, 282
rhomboid, 54, 76, 89, 92, 93, 95, 247, 261
rotatores spinae, 54, 57, 58, 59
sartorius, 21, 30, 36, 39
scalenus anticus, 62, 67, 76, 82, 83
scalenus medius, 62, 67, 76, 82
scalenus posticus, 62, 67, 76, 82
semimembranosis, 21, 30, 36, 39, 42
semispinalis colli, 82, 83
semispinalis dorsi, 54, 57, 76, 82
semitendinosis, 21, 30, 36, 39, 42
serratus magnus, 62, 67, 89, 94, 249, 265
serratus posticus inferior, 54, 62
serratus posticus superior, 54, 62, 66, 76
soleus, 21, 22
splenus capitis, 54, 76, 79, 82, 83, 84
splenus colli, 54, 55
spinalis dorsi, 54, 55
sterno-hyoid, 62
sterno-mastoid, 62, 79, 82, 83, 84, 87, 92
subclavius, 62, 87, 92, 93
subscapularis, 87, 90, 97, 98
supinator brevis, 91, 100, 104

supinator longus, 91, 251
supraspinatus, 87, 90, 97
tendon of Achilles, 25, 142
tensor fascia lata, 39
tensor vaginae femoris, 36
teres major, 89, 91, 97, 98
teres minor, 89, 91
tibialis anticus, 18, 21, 23, 140, 256
tibialis posticus, 18, 21, 22, 256
trachelo-mastoid, 54, 76, 79, 82, 83, 84
transversalis, 36, 48
transversus pedis, 18, 141
trapezinus, 54, 76, 79, 82, 84, 87, 92, 93, 94, 116, 247, 249, 251, 259, 261, 265, 266, 271, 277
triangularis sternae, 62
triceps, 89, 91, 97, 99, 101, 102, 118, 249, 251, 267, 271, 277, 282
vastus externus, 27
vastus internus, 27

Naismith, 253
Neck, 74
  joints of, 80
  ligaments of, 75
Neural arch
  of cervical spine, 74
  of lumbar spine, 45
  of thoracic spine, 53
Nicholls and Mallam, 260
Normal posture (see Posture, normal)
Nuchal lines, 78

Obliquus capitus inferior muscle, 76, 83
Obliquus capitus superior muscle, 76, 79, 84
Obliquus externus muscle, 49, 51, 58, 59, 62
Obliquus internus muscle, 36, 49, 51, 58, 59, 62, 73
Obturetor externus muscle, 27, 36, 40, 57
Obturetor internus muscle, 27, 36, 39, 40, 57
Occipito-atlantoid joint, 81
Occipito-cervical ligament, 76, 79, 81
Occipito-odontoid joint, 81
Occiput, 76, 77

Odontoid process, 75, 80, 81
Olecranon fossa, 91, 102
Olecranon process, 91, 99, 102
On hands and knees position, 9, 129
 defined, 10
 fundamental gymnastic, 129
Opponens minimi digiti muscle, 105, 106
Opponens pollicis muscle, 105, 108, 110
Orbicular ligament, 99, 100, 101, 102
Orders of levers, 10
 defined, 10
Os innominatuon, 35, 38
Os magnum, 104

Padding and strapping of foot, 148
Palmar ligament, 105
Palmaris longus muscle, 91, 101, 107
Patella, 29
Pathological deviations from normal posture, 136
Pectineus muscle, 36
Pectoralis major muscle, 62, 67, 72, 87, 89, 91, 93, 94, 95, 96, 97, 98, 116, 247, 249, 251, 265, 266, 267, 271, 276, 277, 282, 283
 compound action of, 116
Pictoralis minor muscle, 67, 92, 93, 94, 247, 249, 276, 277, 282, 283
Pedicles
 of cervical spine, 75
 of lumbar spine, 45, 46
 of thoracic spine, 53
Pelvic inclinometer, 174
Pelvis, 27, 35, 38, 256, 258, 265
 bones of, 35, 85
 defined, 35
 ligaments of, 36
 muscles of, 36
Peronei muscles
 brevis, 18, 19, 21, 23, 141
 longus, 18, 19, 21, 23, 141
 tertius, 18, 19, 21, 23, 141
Petrous portion, 78
Phlangeal joint, 85
Phalanges, 104, 106, 108, 109, 110
Pisiform bone, 104
Pitcher, in base ball, 240
Plantar arch of foot, 15

Plantar ligament, 141
Plantares muscle, 27
Plaster of Paris jackets in treatment of scoliosis, 208
Pole raulting, 278
 kinesiology of, 279
Pollaiuolo, 121
Popliteus muscle, 21, 27
Positions, fundamental (see Fundamental positions)
Posterior atlanto-axoidean ligament, 75, 81
Posterior chondro-sternal ligament, 61, 65
Posterior common ligament, 36, 47, 48, 51, 54, 75, 80
 of the spine, 36
Posterior costo-transverse ligament, 54, 64
Posterior crucial ligament of the knee, 30
Posterior ligament,
 of ankle joint, 22
 of elbow, 101
Posterior occipito-atlantal ligament, 76, 79, 81
Posterior radio-ulnar ligament, 100
Posture, faulty
 Exercises, 180, 181, 183, 184, 185, 186 187, 190, 191
 functional lateral curvature, 167, 170
 gorilla type, 32, 167, 177
 kangaroo type, 166, 177
 measurement and recording, 172, 174
 muscle development, posture training, 178
 pathological deviations from normal, 136
 posture training, 194, 195, 196
 preventive measures, 180
 summary, 198
 treatment of, 177, 178, 179
Posture in statuary, 120
Posture, normal, 167
 analyzed, 167
Poupart's ligament, 36
Preventive measures for foot disabilities, 156

INDEX

Pronation, 103
Pronator quadratus muscle, 99, 100, 103, 249, 251
Pronator radii teres muscle, 91, 99, 100, 101, 103, 249, 251
Prone lying position, 9, 130
  defined, 10
  fundamental gymnastic, 130
Protection in relation to muscle re-education, 6
Proximal and distal, named, 109
Psoas magnus muscle, 36, 42, 48, 49, 51
Putting the shot, 285
Pyriformis muscle, 27, 36, 40

Quadratus lumborum muscle, 36, 48, 49, 51, 57, 58, 59, 62, 274
Quadriceps femoris muscle, 21, 27, 30, 36, 39, 40
Quadriceps muscle, 243, 244, 245, 246, 247, 252, 265, 266, 274, 276, 277, 282

Race walking, 246
  kinesiology of, 246
Radio-ulnar joints, 102
  ligaments of, 103
  movements of, 103
  muscles of, 103
  supination of, 103
Radius bone, 92, 98, 99, 100, 101, 102, 104, 107
Ramus, 79
Ratios of foot, 142
Reciprocal action of muscles passing over more than one joint, 42
Reciprocal movements and leverage problems, involving both hips and both ankles, 43
Recording of scoliosis, 205
Rectus abdominis muscle, 49, 51, 57, 62
Rectus capitus anticus major muscle, 76, 79, 82, 83, 84
Rectus capitus anticus minor muscle, 76, 79
Rectus capitus lateralis muscle, 76, 79, 84
Rectus femoris muscle, 36, 282
Respiratory expiration, 63, 69
Respiratory inspiration, 62

Rhomboid ligament, 87, 92, 93, 97
Rhomboid muscles, 54, 76, 89, 92, 93, 95, 247, 261
Rib, 53, 60, 61
  costo-chondral, 60
  depression and inversion, 63, 65
  elevation and eversion, 63, 65
  false, 60
  floating, 60
  head, 61
  joints of, 62
  kinesiology of, 69
  leverage of, 69
  neck, 61, 64
  shaft, 61
  true, 60
  tubercle of, 64
Right handedness, theories of origin, 240
Rotary lateral curvature of spine, 70, 199
  braces and corsets, 210
  classwork gymnastics, 233
  corrective exercises for, 211, 214, 216, 218, 221, 223, 225, 226, 227, 228
  definition and mechanism, 199
  estimate of treatment, 234
  "idiopathic" type, 199
  measurement of, 201, 203
  operations for, 233
  plaster of Paris jackets, 208
  posture training, 23
  preliminary exercises, 215
  recording of, 205
  summary of, 236
  treatment of, 207, 208
  x-ray examination, 201
Rotation of chest and upper spine
  Forbes-Truslow explanation of complicated mechanism, 70
Rotatores spinae muscle, 54, 57, 58, 59
Row boating, 254
Rowing, 241
Running, 240, 241, 253
Running high jump, 254, 284

Sacro-iliac synchondroses, 167
Sacro-lumbar joint, 48
  kinesiology of, 50
  ligaments of, 48

310 INDEX

Sacro-lumbar joint, movements of, 49
  muscles of, 49
Sacro-sciatic ligaments
  greater, 36
  lesser, 36
Sartorius muscle, 21, 30, 36, 39
Scalenus anticus muscle, 62, 67, 76, 82, 83
Scalenus medius muscle, 62, 67, 76, 82
Scalenus posticus muscle, 62, 67, 76, 82
Scaphoid bone, 104
Scapula, 86, 87, 89, 96, 258, 259
  spine of, 87
Scapulo-clavicular muscle, 92
Scoliosis, 73, 199
  braces and corsets, 210
  classwork gymnastics, 233
  corrective exercises for, 211, 214, 216, 218, 221, 223, 225, 226, 227, 228
  definition and mechanism, 199
  estimate of treatment, 234
  "idiopathic" type, 199
  measurement of, 201, 203
  operations for, 233
  plaster of Paris jackets, 208
  posture training, 231
  preliminary, 215
  recording of, 205
  summary of, 236
  treatment of, 207, 208
  x-ray examination, 201
Semilunar bone, 104
Semi-lunar cartilages of the knee, 29
Semimembranosis muscle, 21, 30, 36, 39, 42
Semispinalis colli, 76, 82, 83
Semispinalis dorsi muscle, 54, 57, 76, 82
Semitendinosis muscle, 21, 30, 36, 39, 42
Serratus magnus muscle, 62, 67, 89, 94, 249, 265
Serratus posticus inferior muscle, 54, 62
Serratus posticus superior muscle, 54, 62, 66, 76
Shoes, 152
Shot-put, 240, 241, 285
Shoulder girdle, 85, 86, 92, 95, 258, 265
Shoulder joint, 85, 86, 92, 259
  kinesiology of, 95
  movements of, 96
  muscles of, 96

Sigmoid cavity, 99, 100, 102
Sitting position, 9, 133
  defined, 10
  fundamental gymnastic, 133
Skating, 240, 253
Skiing, 253
Skull, 76, 77
Skulling, 254
Soleus muscle, 21, 22
Spastic flat foot, 163
  treatment for, 163
Spinalis dorsi muscle, 54, 55
Spine, 70, 73, 74, 167, 170, 171, 258
  cervical, 74
  functional lateral curvature of, 167, 170, 171
  rotary lateral curvature of, 70
  scoliosis, 73
Spinous process
  of cervical spine, 75
  of thoracic spine, 53
Splenius capitus muscle, 54, 76, 79, 82, 83, 84
Splenius colli muscle, 54, 55, 76, 82, 83
Spring sitting position, 9, 134
  defined, 10
  fundamental gymnastic, 134
Standing position, 9
  defined, 10
Statuary, posture in, 120
Stellate ligament, 54, 61
Sterno-clavicular joint, 85, 92, 94, 258
Sterno-clavicular ligament, 87, 92, 93
Sterno hyoid muscle, 62
Sterno-mastoid muscle, 62, 79, 82, 83, 84, 87, 92
Sternum, 61, 258
  bones of, 86, 92
Strapping and padding of foot, 148
Styloid process, 78, 99, 100
Subclavius muscle, 62, 87, 92, 93
Subscapularis muscle, 87, 90, 97, 98
Superior chondro-sternal ligament, 61, 65
Superior costo-transverse ligament, 54, 61, 64
Superior radio-ulnar joints, 85, 98, 101, 102
Supinator brevis muscle, 91, 100, 104
Supinator longus muscle, 91, 251

Supports for foot, 152
Supraspinatus muscle, 87, 90, 97
Supraspinous ligament, 36, 48, 54, 75, 80
Sustentaculum tali, 22
Swimming, 241, 248
   breast stroke, 249
   crawl, 252
   frog kick, 251
   kinesiology of breast stroke, 249
   on back, 252
Symphysis pubis, 35

Temporal bone, 78
Tendo Achillis, 25, 142
Tennis, 240, 241, 272
Tensor fosciae latae, 39
Tensor vaginae femoris muscle, 36
Teres major muscle, 89, 91, 97, 98
Teres minor muscle, 89, 91
Thigh, bones of, 27, 85
Thoracic joint, 56
   kinesiology of, 58
   leverage of, 59
   movements of, 57
   muscles of, 57
Thoracic spine, 53
   articular process of, 53
   bones of, 85
   lamina of, 53
   ligaments of, 54
   muscles of, 54
   neural arch of, 53
   pedicles of, 53
   spinous process of, 53
   transverse process of, 53
Thorax, 62
   bones of, 93
   joints of, 63
   movements of, 62
   muscles of, 62
   respiratory movements of, 65
Throwing the discus, 283
Thumb, 105, 107, 108, 110
   joints of, 110
Tibia, 21, 256
Tibialis anticus muscle, 18, 21, 23, 146, 256
Tibialis posticus muscle, 18, 21, 22, 256
Tidal air, 67

Trachelo-mastoid muscle, 54, 76, 79, 82, 83, 84
Transversalis muscle, 36, 48
Transverse ligament, 27
   of the hip joint, 36, 38
   of the neck, 75, 81
Transverse process
   of cervical spine, 75
   of lumbar spine, 45
   of thoracic spine, 53
Transversus pedis muscle, 18, 141
Trapezium bone, 104, 105, 107
Trapezius muscle, 54, 76, 79, 82, 84, 87, 92, 93, 94, 116, 247, 249, 251, 259, 261, 265, 266, 271, 277
   compound action of, 116
Trapezoid bone, 104
Transverse ligament, 105
Trapezoid ligament, 87
Treatment of scoliosis, 207
   braces and corsets, 210
   corrective exercises for, 211
   general outline of, 208
   plaster of Paris jackets, 208
Triangularis sternae muscle, 62
Triceps muscle, 89, 91, 97, 99, 101, 102, 118, 249, 251, 267, 271, 277, 282
   compound action of, 118
Trachlea, 92
Tuberosities of humercus, 90
   greater, 90, 96
   lesser, 90
Types of faulty posture, 167
   functional lateral curvature, 167, 170
   gorilla, 32, 167, 177
   kangaroo, 167, 177

Ulna bone, 89, 91, 98, 99, 100, 101, 102, 104, 105, 107
Unciform bone, 104
Upper extremity, 85
   kinesiology of, 111

Vastus externus muscle, 27
Vastus internus muscle, 27

Walking, 241, 242
   kinesiology of, 242
   race, 246

Wallace, Lew, 255
Weak foot, 138
  anatomy of, 138
  examination, measurement and recording of, 141, 142, 143, 146
  exercises, 157, 158, 159, 160, 161, 162
  hallux valgus, 141
  kinesiology of, 138
  metatarsalgia, 138, 141
  metatarsus primus varus, 141
  padding and strapping, 148
  ratios, 142
  supports, 152
  treatment of, 148, 165
Wrist, 85, 104
  compound action of muscles, 119
  joints of, 99, 106, 107
  movements of, 106
  muscles of, 106

Xiphoid appendix, 61

Date Due

WITHDRAWN